CHINA

CHINA

Towards Sustainable Industrial Growth

INDUSTRIAL DEVELOPMENT REVIEW SERIES

Published by Blackwell Publishers for the

UNITED NATIONS INDUSTRIAL DEVELOPMENT ORGANIZATION

First published 1992

ISBN 0-631-17525-3 ISSN 0956-6686

Blackwell Publishers
108 Cowley Road, Oxford, OX4 1JF, UK

Blackwell Publishers Inc.
3 Cambridge Center,
Cambridge, MA 02142, USA

British Library Cataloguing in Publication Data
China: towards sustainable industrial growth – Industrial development review series, 0956-6686; 5
1. China (People's Republic of). Industries
I. United Nations *Industrial Development Organization*
338.0954

ISBN 0-631-17525-3

Acknowledgements

This Review has been prepared by the United Nations Industrial Development Organization under the general direction of Torben M. Roepstorff, Senior Industrial Development Officer, in collaboration with UNIDO consultant Georgina Wilde, whose contribution is acknowledged. The preparation of this Review was greatly facilitated by co-operation with Huang Fan-Zhang, Director, Economic Research Centre of State Planning Commission, Beijing, Assisted by Zhao Longyue and Han Xiaojin, and Gao Yuxian, Deputy General Manager, China Statistical Information and Consultancy Service Centre, Beijing. Substantial Contributions were also made by Jebamalai Vinanchiarachi, UNIDO Industrial Development Officer, and UNIDO consultant Wolfgang Pelz. Overall guidance was provided by Herman Muegge, Head of the Regional and Country Studies Branch of UNIDO.

Typeset by Information Press, Oxford
Printed and Bound in Great Britain by The Alden Press, Oxford

CONTENTS

LIST OF TABLES, FIGURES AND BOXES

TABLES

FIGURES

BOXES

CHAPTER II

PREFACE

This Industrial Development Review of China is part of a sales series aimed at strengthening the "country focus" of UNIDO activities. Within the framework of the work programme of the Regional and Country Studies Branch of UNIDO which monitors the industrialization process, the Reviews provide a general survey and brief analysis of each country's industrial development process. The Reviews are intended to provide a service to those within UNIDO and other international agencies concerned with industrial policy, planning, project development and implementation, and to be a ready source of information for governments, investors, industrialists, entrepreneurs, policy-makers, international organizations, aid agencies, academics, and research institutes.

The Reviews have two separate but interrelated objectives: they are designed to facilitate and promote the activities of UNIDO, as well as to serve as an informative and analytical document for the international industrial community. It is known from experience that readily available reference material on the industrial sector is eagerly sought. The favourable responses received from regular readers both inside and outside UNIDO have facilitated extension of the scope of the Reviews in successive issues.

The scope and dimensions of the sales series of the Reviews are designed to accommodate the needs of a wide readership in the international industrial community associated with industry, finance, trade, business, research and government. The Reviews aim at providing a basis for undertaking in-depth analyses of specific aspects of industrial policies, strategies and programmes and at providing a basis for informed discussion of industrial development trends and policies.

The Reviews are also intended to strengthen the Organization's relationship with the private sector. By acquiring a wide readership for this series, UNIDO hopes to provide new and pertinent information on the role of industry: information that is essential to understanding and accelerating the process of industrialization.

This Review comprises four Chapters. Chapter I presents an overview of the Chinese economy and analyses the economic environment with a focus on the outlook for the 1990s. The policy framework for industrial development and the investment environment are examined in Chapter II. The structure and performance of the manufacturing sector are examined in Chapter III with particular reference to value added, output, employment, productivity, wages and salaries, international trade, investment and financing patterns, industrial location and regional development perspectives. Finally, Chapter IV analyses the retrospects and prospects of key industrial branches. Data on economic and industrial trends are presented in Annex A. Annex B contains technical notes on Chinese statistical terms and concepts. The Review also furnishes information pertaining to industrial investment procedures, opportunities, institutions and subsector-specific consulting firms in a set of Annexes.

This Review is based on information available as at 31 July 1991.

EXPLANATORY NOTES

References to dollars ($) are to United States dollars, unless otherwise stated.

Chinese currency, Renminbi (Rmb), is denominated in Yuan, i.e., one Renminbi denotes one Yuan. One Yuan signifies 10 jiao (=100 fen).

Dates divided by a hyphen (1990-1991) indicate the full period, including the beginning and the end years.

References to the Federal Republic of Germany and the German Democratic Republic indicate the period prior to unification of the two German States, on 3 October 1990. As of that date, the designation "Germany" is used. In tables and listings, the former component States are listed under "G": Germany, Federal Republic of; German Democratic Republic.

In Tables: Totals may not add precisely because of rounding. Two dots (..) indicate that data are not available or not separately reported. A dash (-) indicates that data are not applicable.

The following *abbreviations* are used in this publication:

BAOSTEEL	Baoshan General Iron and Steel Works
CNAIC	China National Automotive Industry Corporation
CONGEDC	China Oil and Natural Gas Exploration and Development Corporation
CMRS	Contract management responsibility system
FAW	First Auto Works
GATT	General Agreement on Tariffs and Trade
GDP	Gross domestic product
GNP	Gross national product
GIOV	Gross industrial output value
GITIC	Guangdong International Trust and Investment Corporation
kW	Kilowatt
MOFERT	Ministry of Foreign Relations and Trade
MFN	Most Favoured Nation
MFA	Multi-Fibre Arrangement
MVA	Manufacturing value added
NIEs	Newly Industrializing Economies
OECD	Organization for Economic Co-operation and Development
PVC	Polyvinyl chloride
R & D	Research and development
RCA	Revealed Comparative Advantage
RSVO	Rural social value of output
SAW	Second Auto Works
SEZ	Special Economic Zone
SINOPEC	Chinese Petrochemical Corporation
sq km	Square kilometers
SOEs	State-owned enterprises
SNA	System of National Accounts
TPC	Taiwan Plastic Corporation
tpd	Tons per day
tpy	Tons per year
TVCEs	Township, village and community enterprises
TVEs	Township and village enterprises
TNCs	Transnational corporations
tcf	Trillion cubic feet
UNIDO	United Nations Industrial Development Organization
YJCCFI	Yizheng Joint Corporation of Chemical Fibre Industry

EXPLANATORY NOTES

BASIC INDICATORS

BASIC INDICATORS I: THE ECONOMY

Gross national product (GNP) (1990, at current prices)	:	Rmb 1,768.6 billion[a/]
Population (end-1990)	:	1,143.3 million
Rural/urban ratio (end-1990)	:	26.4/73.6
Population density (end-1990)	:	119.1 persons per square kilometre
Annual average growth of population	:	2.3 per cent (1960-1978) 1.4 per cent (1979-1990)
Labour force (end-1990)	:	567.4 million
Adult literacy rate (1990)	:	84 per cent
Life expectancy (1982)	:	67.9 years
GNP per capita (1990)	:	Rmb 1,558
Annual average growth rate of GNP per capita (1979-1990)	:	7.2 per cent

Average annual growth rate of GNP (Percentage) :

1979-1990	1984	1985	1986	1987	1988	1989	1990
8.7	14.6	12.7	8.3	11.0	10.8	3.9	5.0

Structure of GNP (at current prices) (Percentage) :

	1978	1985	1988	1989	1990
Primary industry	28.4	29.7	27.3	26.6	28.4
Secondary industry	48.6	45.2	47.0	45.7	44.3
Tertiary industry	23.0	24.8	25.7	27.7	27.3

Inflation (Percentage) :

1985	1986	1987	1988	1989	1990
8.8	6.0	7.3	18.5	17.8	2.1

Balance of foreign trade (trade surplus) (1990)	:	$8.74 billion
Gold reserves (1990)	:	12.7 million oz
Foreign exchange reserves (1990)	:	$28.6 billion
Total foreign debt (end-1990)	:	$52.5 billion

Debt service ratio (Percentage) :

1985	1990
6.3	9.2

Rate of growth of consumer prices	:	6.9 per cent per annum (1979-1990)
Rate of real growth of money supply	:	1.3 per cent per annum (1990)

Currency exchange rate (Yuan equivalents to $1)[a/] :

Early 1984	May 1985	1988	1989	1990	July 1991
2	2.8	3.73	3.78	4.78	5.3

a/ The Chinese currency is called Renminbi (Rmb) and is denominated in Yuan. One Renminbi denotes one Yuan. One Yuan signifies 10 jiao (=100 fen).

BASIC INDICATORS II: RESOURCES AND TRANSPORT INFRASTRUCTURE

Resources		
Land area:	:	9.6 million square kilometres
Agricultural production (1990) (Million tons)	:	Grain (446.2 million), cotton (4.51 million), oilseeds (16.1 million), sugar cane (57.6 million)
Forest (1990)	:	124.7 million hectares (1.87 billion mu)
Mineral resources (1990)	:	Coal reserves (954.4 billion tons), iron ore reserves (50.1 billion tons) (largely unexplored)
Energy production (1990)	:	Coal (1,080 million tons), crude oil (138.3 million tons), natural gas (16.0 billion cubic metres)
Rate of growth of energy production (annual average per cent, 1979-1990)	:	Coal (4.8 per cent), crude oil (2.4 per cent), natural gas (0.9 per cent)
Energy production shares (1990) (Percentage)	:	Coal (74.1 per cent), crude oil (19.0 per cent), hydropower (4.9 per cent), natural gas (2.0 per cent)
Transport		
Roads (1990)	:	1,028,300 kilometres
Railway track (1990)	:	53,378 kilometres
Inland waterways (1990)	:	109,200 kilometres
Civil aviation routes (1990)	:	506,700 kilometres

BASIC INDICATORS III: THE MANUFACTURING SECTOR

Manufacturing net output (Rmb 100 million) (1990)	:	4,541.27 (at current prices, by distribution methods)
Manufacturing net output per capita (1990)	:	Rmb 397.21
Employment in manufacturing as	:	121.58 million persons
percentage of total labour force (1990)	:	21.4 per cent
Composition of net industrial output value (Percentage) (1990)	:	

	1978	1982	1985	1988	1989
Light industry	37.6	43.5	39.3	42.4	42.2
Heavy industry	62.4	56.5	60.7	57.6	57.8

	1990
Light industry	43.4
Heavy industry	56.6

Net output per manufacturing worker (1990)	:	Rmb 3,735
Average annual growth rate of net industrial output value (Percentage)	:	

1966-1970	1971-1975	1976-1980	1981-1985
12.3	9.0	9.2	10.1

1986	1987	1988	1989	1990
9.6	14.3	20.8	8.5	15.8

Trade in manufactures (1990)	:	
Total value - exports	:	Rmb 46.21 billion
Total value - imports	:	Rmb 43.49 billion
Share of manufactures in (1990)	:	
Total exports	:	74.4 per cent
Total imports	:	81.5 per cent

BASIC INDICATORS IV: INTER-COUNTRY COMPARISON OF SELECTED INDICATORS

Indicator	Unit	People's Republic of China	Brazil	India	Indonesia	Republic of Korea
Population (mid-1989)	Million	1,113.9	147.3	832.5	178.2	42.4
Area	'000 sq km	9,561	8,512	3,288	1,905	99
GNP per capita (1989)	$	350	2,540	340	500	4,400
Average annual rate of inflation (1980-1989)	Per cent	5.8	227.8	7.7	8.3	5.0
Private consumption (1989)	Per cent of GDP	59	65	67	53	52
Gross domestic investment (1989)	Per cent of GDP	24	22	24	34	35
Gross domestic savings (1988)	Per cent of GDP	21	26	21	37	37
Exports of goods and services (1989)	Per cent of GDP	14	7	8	26	34
Energy consumption per capita (1989)	kg of oil equivalent	591	897	226	263	1,832
Food industry (1988)	Per cent of MVA	12	14	10	26	11
Textile and clothing (1988)	Per cent of MVA	14	10	13	13	15
Machinery and transport equipment (1988)	Per cent of MVA	25	21	27	3	32
Chemicals (1988)	Per cent of MVA	11	13	17	9	9
Other industries (1988)	Per cent of MVA	38	42	33	49	33
Manufactured exports to OECD countries (1989)	$ million	26,778	11,674	8,125	4,612	42,601
Balance of payments[a/] (1989)	$ million	-4,530	-1,040	-7,538	-1,368	5,056
Gross international reserves (1989)	$ million	23,053	10,505	8,048	6,444	15,342
External debt (1989)	$ million	44,857	111,290	62,509	53,111	33,111
Debt service ratio[b/] (1989)	Per cent	5.5	15.5	14.3	14.9	3.8

Source: World Bank, *World Development Report 1991* (Washington D.C. 1991).

a/ Current account balance after official transfers.
b/ Total interest payments on long-term debt as percentage of exports of goods and services.

SUMMARY

The rapid growth of the Chinese economy since the introduction of structural reforms in the early 1980s slowed markedly after showing apparent signs of overheating in 1988. The potential of the economy to reach new heights was revealed during the first three years of the Seventh Plan (1986-1990), when real GNP grew at or close to double-digit rates. The pace of economic expansion eased to 3.9 per cent in 1989 and 5.0 per cent in 1990. Although the overall target of the Seventh Plan was surpassed, macroeconomic imbalances surfaced during 1989-1990.

A slowdown in the rate of expansion of the agricultural sector in 1988 resulted in a fall in grain output and static staple supplies, putting upward pressure on prices. Serious overheating of the economy in 1988 was exacerbated by the over-rapid expansion of industrial output and sharply rising consumer and investment demand. Coupled with higher food prices, this led to an outburst of inflation. The inflation, which was partly the result of price decontrol, was exacerbated by substantial wage and bonus awards to workers by enterprises. The external sector of the economy was also giving cause for concern as soaring imports resulted in sizeable current account deficits in 1988 and 1989.

An austerity drive in 1989 was aimed at, among other things, wholesale cancellation of projects, reforming the management of the key State-owned enterprises, stringent controls on imports, allocation of scarce industrial inputs and a measured pace of economic expansion. Good harvests in 1989 and 1990 helped restore stability, while the curb on investment and the very slow growth of industrial output in the second half of 1989 and the first quarter of 1990 eased pressure on scarce resources. The devaluation of the currency against the dollar - from Rmb 3.73 to Rmb 4.72 in December 1989 and again to Rmb 5.3 a year later - and import controls helped bring about a dramatic turnaround on the trade balance in 1990. The fear of a full-scale recession with an adverse impact on employment compelled the government in 1990 to stimulate demand by easing the high interest rates, tight credit and strict investment controls. A 7.6 per cent increase in agricultural output and a modest (by Chinese standards) 7.6 per cent growth of industrial production pushed the growth of real GNP from 3.9 per cent in 1989 to 5.0 per cent in 1990.

As China enters its Eighth Plan (1991-1995), macroeconomic stability has been restored and, most importantly, inflationary pressures have been brought under control without sacrificing growth. The decline in the growth of agricultural output has been reversed, and the external balance is healthy. However, the threat of renewed inflation, inefficient State sector, chronic budgetary deficit, the high rate of growth of the population, infrastructural bottle-necks and acute scarcity of industrial inputs are posing challenges to rapid expansion. A modest annual economic growth of 6 per cent is envisaged for the 1990s. It seems likely that attempts will be made to keep the growth of gross industrial output below 10 per cent a year in order to straighten out the imbalances.

Decades of massive industrialization have brought China to the front rank in the global industrial map. China's share of world industrial output more than doubled during 1977-1986, and by the end of 1987 manufacturing value added ranked seventh in the world. The vast majority of industrial activity (54.5 per cent in gross industrial output value in 1990) is still concentrated in the State-owned sector, either controlled by Central Ministries or run by provincial governments. In 1988, of the total 8.1 million registered enterprises, 99,100 were State-owned large- and medium-sized enterprises. State-owned units employed 18 per cent of the labour force in the same year. There are also thousands of much smaller-scale plants using locally produced machinery and run under local control, as well as the large numbers of collective and private industries which have sprung up since 1979. These have traditionally been encouraged as a means of absorbing the growing rural labour force and have multiplied as a result of the encouragement of more diverse ownership and management structures during the reform decade. The number of enterprises owned by rural or urban collectives, or by individual families or a combination of the two, and the number of concerns involving foreign investment has greatly increased.

The outstanding features of the double-digit growth of Chinese industry in the preceding decades had been structural changes that occurred in distinct phases of development. During 1957-1978, the share of consumer goods in gross industrial output declined and that of heavy industry rose significantly. The period 1978-1983 saw a revival of structural trends reflecting the increased emphasis policy-makers placed on the expansion of light industries to keep pace with the rising consumer demand. However, during 1983-1989, heavy industry regained much of the lost ground with an increase in its share of the net industrial output from 54.1 per cent in 1981 to 57.8 per cent in 1989, but fell marginally to 56.6 per cent in 1990. The period 1983-1988 was one of exceptionally brisk expansion with an accelerated pace of investment growth, when gross industrial output value grew at an average annual rate of 18 per cent. In the wake of rectification policy measures, State-owned enterprises reduced investment by 9.2 per cent, collective enterprises by 28.1 per cent and private enterprises by 4.3 per cent in 1989.

The efficiency of Chinese industry in general has been impaired by a variety of structural factors. These include the ownership structure, the system of production according to quotas set by the State, the lack of financial and managerial autonomy, a pricing system which fails to reflect relative scarcities, a shortage of key inputs, overmanning on a large scale in accordance with socialist values, a poorly developed internal market and a shortage of skills. The practice of reselling scarce raw materials and intermediate inputs allocated to the planned sector fetched easy profits for the State enterprises and State businesses in 1990. As the State monopoly was still dominant in the underdeveloped market mechanism, rising raw material prices failed to evoke an adequate supply response. The incremental capital-output ratio in China is both higher than in other economies at a similar stage of development and rising, despite a diversified product profile and considerable improvement in the quality of industrial plant over the last ten years.

China invested heavily in industrial modernization in the 1980s in order to enhance productivity through technical change, economies of scale and efficient allocation of resources. Against a slow increase in the overall labour productivity until the early 1980s, the growth of labour productivity in real terms was significant in selected segments of industry, particularly in electronic apparatus, consumer electronic equipment, electric equipment and machinery, transport equipment, plastic products, arts and crafts articles, furniture manufacturing, etc. In 1988, industrial wages rose by 28.1 per cent in nominal terms. Adjusting for the official inflation rate of 18.5 per cent, wages rose by 9.6 percentage points in real terms. The growth of labour productivity in the above subsectors of manufacturing was faster than the increase in real wages in 1988. In 1990, cotton textiles and power production and supply recorded improvements in labour productivity. Around 80 per cent of the State-owned enterprises are 15 to 80 per cent overstaffed, and 15-20 million industrial workers are in need of reassignment. This calls for prudent manpower planning within the industrial sector.

A striking feature of industrial expansion since 1984 has been an increased reorientation towards the market economy, as enterprise altered investment and production patterns in order to respond effectively towards rapidly changing demand patterns. Self-financing became an important source of investment for all categories of enterprise. It is estimated that during the second phase of the reform period (1984-1988) even State enterprises financed 20 per cent of fixed capital formation and about one-third of modernization through retained profits. Self-financing as a source of investment finance rose significantly in 1988 and 1989. In 1990, it stood at 52.4 per cent of investment finance

There is much scepticism about profiteering through the resale of raw materials and scarce commodities. Revenue from the sales of products of State-owned enterprises rose by 275 percentage points during 1982-1989. The proliferation of profiteering and other speculative activities led to distortions of the market mechanism. An investigation into the financial performance of State-owned enterprises uncovered Rmb 1.13 billion of illegal gains in 1989. Due to public discontent, the government initiated several rounds of cleanup and rectification of profiteering activities. By June 1989, a total number of 6,481 State-owned enterprises were dissolved or merged, and 10,386 separated from their founding Party and government departments.

In cleaning up and rectifying companies the government endeavours to create a better environment and conditions for reform and openness.

Efficiency levels have also improved since the adoption of measures by the State Council to rationalize production structures and to construct industrial conglomerates, with a view to reducing production duplication and to benefiting from economies of scale. However, this has led to an increase in the level of industrial concentration and to the emergence of about 20 giant corporations which operate in key industrial sectors and are in a position to influence their price and output level decisions on a wide range of markets.

Export diversification too was particularly rapid during the 1970s and 1980s. The overwhelming evidence from the statistical estimates prepared by UNIDO for the early 1980s shows that China has increased its comparative advantage in natural resource-based industrial products, lost some ground in labour-intensive products (particularly textiles) and failed to make any significant headway in technology- and capital-intensive export markets. The country's comparative advantage seems to be in products which have experienced declining shares in world trade. China has the lowest comparative advantage in capital- and technology-intensive products which are rapidly increasing their share in world trade.

Products of light industry dominate the country's export profile. In dollar terms, canned food, silk and satin material, fur shoes, sewing machines, bicycles and machine tools grew significantly on the export front in 1990. In the 1980s, export growth was stimulated when China's effective exchange rate was depreciated by over 100 per cent. Around 80 per cent of China's imports comprise raw materials, industrial machinery and transport equipment. It appears that the recent cutbacks in imports and the go-slow policy are a short-term reaction to the overheated economy. The United States is most likely to extend the Most Favoured Nation (MFN) status to China. It has become increasingly important for China to accept positive reciprocity in trade.

The relatively slow growth of the food processing industry in most of the 1980s at 7 per cent per annum, about half of the manufacturing average, was due to the large-scale decline in grain production, lack of diversification in product mix and inadequate packaging. The average food manufacturing firm is too small to exploit economies of scale. Although some progress has been achieved in the modernization of larger firms, over a third of the equipment in use in the 1980s was of the 1940s vintage. Modernization investment in food manufacturing is inadequate to meet capital replacement requirements. Horizontal integration and production rationalization is hampered by provincial and local concern in achieving and maintaining food self-sufficiency. Aquatic products rank first in China's food exports, followed by canned food, tea and meat products. China is the world's third largest exporter of tea after India and Sri Lanka. Special tea ingredients are more readily available for project development specialists to formulate distinctive tea-based beverages. The development of the packaging industry is a prerequisite for the rapid growth of exports of food manufactures. This requires an expansion of the domestic production of quality paper, plastics and glass and also a standardization of the production procedures for cans and bottles. There is considerable scope for foreign firms in expanding and upgrading the domestic production capacity of the food packaging branch.

It is estimated that by the year 2000, domestic consumption of meat, poultry, eggs and fish will have doubled, and the projected grain output will be far below demand. It is imperative to gradually develop a food consumption structure and make the supply response well-matched with the structure. With a view to achieving this, priorities are accorded to selected segments of the food industry. In cereal and oil processing, the rice- and flour-producing segments are destined for expansion. The quality and varieties of wheat flour will also be increased. Oil refining, purifying, decolourizing and deacidifying will be upgraded. Special oils for food, such as hydrogenated oils, margarine and shortening, are to be developed. Efforts are being directed towards large-scale manufacture of products based on beet and cane sugar. Attention is also focused on combining coarse grains with refined grains to produce foods of reasonable nutritional value appropriate to China's resource endowment and pattern of food consumption.

China's textile and clothing industry has made significant headway in production and exports. China ranks first in cotton yarn export and sixth in garment exports in the world. One-fourth of China's total foreign exchange earnings stems from textile and garment exports. Against a significant increase in production and exports, the industry's overall performance has slowed down relatively. Beginning in 1982, the textile industry started to produce a growing unsaleable surplus, with declining productivity (until 1988). This revealed the pressing need for equipment modernization in order to achieve efficiency and to turn out higher quality products. Much of the idle capacity in the industry comes from the shortage of cotton. In the industrial heartland of Shanghai, actual delivery of cotton in 1990 was around 36 per cent of the planned allotment and 26 per cent of the amount needed to keep the installed spindles and looms running. This resulted in substantial underutilization of cotton spinning capacity, and 7,000 persons out of work in the second half of 1990. This situation can be generalized to textile enterprises located in 12 Chinese coastal cities. Lack of working capital, cash flow overdue and soaring costs of raw materials together have led to a severe burden on the textile industry in China. In order to alleviate the shortage of raw materials, efforts are under way to raise the proportion of chemical fibres and the use of non-cotton fibres, such as blended fabric and interwoven products. On the export front, China aims at increasing the annual value of silk exports from $1,650 million in 1988 to $2,500 million in the first half of the 1990s. China's textile and clothing export target for the year 2000 is set at $20 billion. Apart from Hong Kong and Japan, the major target markets continue to be the United States and the EC. The Multi-Fibre Arrangement (MFA) restrictions do not seem to have prevented the rapid growth of exports. This was partly due to the price competitiveness of non-quota items that have grown significantly in recent years. China has constantly made full utilization of quota imposed by the United States, while a good number of textile products gained improved access to the EC in recent years due partly to quota overruns in 1986 and 1987. While China continues to reap the benefits of relatively low prices of its non-quota items, restricted items will need to be specifically identified as target product areas for substantial improvements in quality. In utilizing foreign investment and advanced technology priority is accorded to: developing the chemical fibre industry in order to increase varieties and/or improve quality through the establishment of high acrylic fibre mills and viscose acetal fibre mills; finishing process in-dyeing and pure cotton fabrics, terylene blends, woollen imitation fabrics, pure cotton knitwear, terylene filaments and shuttle woven fabrics in order to keep pace with changing patterns; advanced textile machinery and testing instruments used in technical transformation of existing enterprises; product design improvements in medium- and high-grade garments; developing new varieties of raw materials and production bases for raw materials; and setting up external sales and information networks.

China abounds in iron ore with 50 billion tons of proven reserves. Although the grade of iron ore deposits is low, it is possible to turn out concentrates with an average iron content of around 65 per cent, against the present 30 per cent iron content, using the country's abundant resources such as high-grade bauxite, magnesite, flint clay and coking coal. China ranks fourth in the world production of iron and steel after the Soviet Union, Japan and the United States. However, almost all iron and steel ore is in short supply. In 1989, China imported 9.13 million tons of steel at a cost of $4.43 billion. The domestic production capacity of steel sheet is 8 million tons per year against an annual demand for 10 million tons, estimated to reach 14-15 million tons in 1995. The production capacity of cold-rolled steel sheet is 2.5 million tons per year, meeting just half the annual demand. The production capacity of hot-rolled steel is also inadequate. Although capacity utilization rates across iron and steel enterprises are relatively high, labour productivity remains lower than in their counterparts in developed countries. However, capital productivity in iron and steel industry has been rising in recent years in the face of increasing use of new technologies. The main emphasis is on improving steel quality and varieties and reducing energy and material consumption in steel production. The slow-down in industrial production failed to strike a balance between demand and supply and there were acute shortages of steel products needed for basic industries such as energy, transportation and raw materials. In order to achieve a balance between supply and demand, China could endeavour to reduce the steel intensity of industrial production and look for alternative means. In rehabilitating existing mills, priority is accorded to the upgrading of large steel mills with an annual production capacity of 1 million tons, special steel plants and plants producing mining equipment. In the production process, emphasis is placed on improvements in steel-making technology and major rolling machines.

While the world chemical industry is faced with weakening demand and overcapacity, raising doubts as to how much of the soaring cost of production can be passed on to consumers given the glut in the world chemical industry and volatile oil prices, the demand for and production capacity of basic chemicals are increasing in China. There has been a constant and steady increase in the output of soda ash and caustic soda over the years. The growth rate of small-scale soda ash enterprises has surpassed that of the major large-scale enterprises in recent years, attributed mainly to improvements in the production level of existing plants. The output of chemical fertilizers (nitrogenous and phosphate) rebounded markedly from severe declines in 1985. By 1990, the production of chemical fertilizers stood at 19 million tons compared with 12 million tons in 1981. China continued to expand the production capacity for high-grade fertilizers, particularly phosphate fertilizer. The newly emerging petrochemical industry has achieved significant strides in capacity expansion and production. However, many petrochemical plants are in need of technical renovation. As most segments of the chemical industry expand from infancy, China continues to remain the world's largest chemical importer. Around 17 per cent of apparent consumption of chemical fertilizers was met by imports in the 1980s. The present demand for high quality pesticides is 25-50 per cent higher than domestic supply in China, which represents one of the largest pesticide markets in the world. Inadequate transport facilities have restrained the optimal use of the country's rich raw material resource base. The prospects for chemical products are linked to energy supplies. The problem of energy shortage will need to be alleviated by energy conservation which needs substantial investment in technical renovation.

China has a diverse resource trove for the production of building materials. There had been a continued steady growth of China's building materials industry during 1979-1987, but its expansion in 1988 was severely affected by the freezing of major construction projects in the larger Chinese towns. In order to offset the decline in domestic demand, China is seeking to expand cement exports. China is one of the leading cement producers in the world. Raw material costs and energy intensities remain high by international standards. Foreign technology is being eagerly sought in order to achieve productivity gains in building materials. Shortages are most serious in the case of plate glass and ceramic fixtures. Utilization of foreign funds for developing non-metallic mineral products for export is envisaged.

The relatively low level of technical efficiency in the Chinese machine tool industry is attributed largely to the lack of a competitive environment. Given the lack of technicians with advanced training in handling modern tools, China's reservoir of human skills in traditional machine tools could be augmented for repairing and maintaining them where the traditional technology is deemed appropriate. Because of the proliferation of a large number of small-scale enterprises with limited ability to fully use modern machine tools, even with the increased accent on modern technology, a large proportion of machine tool production is likely to continue on traditional lines.

China has actively encouraged joint venture collaboration within the automobile industry. Despite the predominance of two major plants, the total number of automobile and motor cycle manufacturers has grown rapidly in recent years, and output value has grown at 20 per cent per annum since 1986. The relatively better financial performance of the automobiles than other segments of the machinery industry reflects the greater level of enterprise autonomy that has been made possible by managerial re-organization, horizontal integration and effective decentralization of the enterprise administrative system. However, capacity utilization remained far below the optimal level in the late 1980s due to a limited degree of specialization and the geographical localized nature of supplier networks. Efforts to form the second car assembly joint venture with Volkswagen of Germany are under way. This new venture, FAW-Volkswagen Automobile Company, is aimed at building 150,000 cars a year by 1996. Sixty per cent of the stake will be owned by the First Automobile Works. The initial equity capital of this new venture is estimated at $405 million. The plan is to assemble Volkswagen Golf-Jetta cars for the domestic market and export, particularly to countries in South-East Asia. However, initial production of around 17,000 cars will be exclusively for the domestic market. Exports are scheduled to commence in 1994.

As part of an earlier licensing and technology transfer agreement with Volkswagen of Germany, the First Automobile Works has been assembling the Audi 100 model car since 1989. Its full capacity is expected to reach 30,000 a year in 1995. The new joint venture is expected to achieve a local content level of 65 per cent in 1994.

China plans to increase automobile production from about 580,000 units annually in 1990 to 1,700,000 units by the year 2000. There is a concern to diversify production and expand regional integration. One of the problems facing the Chinese automobile industry is the growing dependence on foreign suppliers and joint venture partners who have been reluctant to source local imputs. Car makers and component suppliers face new developments in the international market due to heightened competition. New challenges stem largely from the build-up of Japanese transplant car production. These trends have far-reaching implications for China.

Lured by the unlimited domestic market, attractive incentives and cheap labour, transnational corporations have formed a number of joint ventures in a range of labour-intensive electronics production activities. China is using market access as a means to put leverage on foreign firms to import technology. IBM, Hewlett Packard, Digital Equipment Corporation, Wang, Hitachi, Fujitsu, Siemens, Olivetti, Unisys and NEC are among other foreign firms which have directed investment into electronics production in China. In 1990, at least six joint ventures with big transnational corporations were formed. IBM and its joint venture partner, the Tianjin Zhonghian Computer Corporation, are planning to manufacture one of the United States's next generation of personal computers. Sales are targeted largely at the domestic market. Joint ventures in electronics are also aimed at capturing the market previously penetrated by Japan, and the Newly Industrializing Economies (NIEs) which are rapidly shifting to skill-intensive products.

China's share of world electronics exports is expected to rise to 2 per cent by the end of this century. Despite the expected increase in electronics exports, China is likely to remain a net importer of electronics in the 1990s. The electronics industry is likely to benefit the most from China's fresh offers of deep tax cuts in the network of industrial parks. China is establishing a country-wide network of industrial parks with a view to attracting foreign and domestic investment into high-tech enterprises. In March 1991, the government approved the legal framework for 27 high-tech industrial parks. These parks are aimed at creating a favourable environment for the development of new technology in order to facilitate the expansion of high technology industries. Foreign joint ventures are completely exempt from income taxes for the first two profit-making years. There are around 2,500 high-tech enterprises operating in these parks. These enterprises turned out $1.3 billion worth of products in 1990. China endeavours to boost the production of high-tech products of industrial parks from Rmb 7 billion in 1990 to Rmb 25 billion in 1996. These parks are expected to play a significant role in bridging the technological gap in microelectronics and electronic information, aerospace, optical electronics, bioengineering, materials science, ecological science and environmental protection, earth science, marine engineering, medical science, biomedical engineering and energy technology. Given the critical role of electronics in enhancing economy-wide competitiveness through the spreading of its benefits across various sectors, the major source of growth will continue to be fuelled by a significant increase in domestic demand. However, a major challenge facing the Chinese electronics industry is to become more competitive internationally. China could endeavour to seize opportunities stemming from global sourcing of components by major players of the global electronics industry who climb up the ladder of high value added products.

There has been all-round exploration and development of the country's energy resources. The share of coal in China's total energy consumption continues to increase, and is expected to reach 76 per cent by the year 2000. Coal production is expected to hit 1.4 billion tons in 2000. Currently, one million tons of coal is turned out by 26 mines in which almost the entire process of coal extraction is mechanized. In order to reduce the country's dependence for coal on the north-west area which has 65.6 per cent of China's coal deposits, there is a tendency to shift coal exploitation to the west. The Ordos coalfields in western China has rich coal deposits. Since the early 1980s, many small local mines have gone into production, and at present about half of the country's coal output is from small mines run by townships, villages and even individuals.

The supply of petroleum products failed to keep pace with demand in the 1980s. In 1990, China's crude production registered a marginal increase of 0.2 per cent over the same period in 1989. As a result, China had to reverse its long-standing policy against imports. With subdued growth of domestic oil production, crude oil supplies have tightened, and since 1988 imports have soared to 110,000 barrels a day in 1989. Exports of crude oil and petroleum products have declined sharply since 1988. The demand for petroleum products is expected to grow between 65-100 per cent by the year 2000, implying a rise of 3.5-4.5 million barrels a day (bpd). In contrast, crude production is expected to rise by only 2.3-3.5 million bpd. Unless efforts to boost exploration and development are intensified, China will need to depend on imports in order to make up this shortfall. China is intensifying efforts to expand the reserve base in order to enhance the production of crude oil to more than one-third higher than present levels. Indications are that the government has planned to spend $20 billion on oil exploration and development during 1991-1995. There are numerous reports of new oil findings. However, new production is hardly keeping pace with rising demand. China's ambitious target of producing 4 million bpd of oil by the year 2000 hinges on the level of foreign investors' participation in oil exploration and development. China's controlled crude price is about 21 per cent of the international price. This is too low to recoup costs and generate cash flow for exploration and development efforts. The country's offshore region is an area of strong interest for foreign investors. Oil agencies plan to accelerate activities in nine oilfields in the South China and Bohai Seas during the Eighth Plan. The China Oil and Natural Gas Exploration and Development Corporation (CONGEDC) is aiming at production targets of 100,000 bpd of crude oil and 116 million metre of gas from offshore fields by the end of 1992. China has around 100 large power stations, each with an installed capacity of over 250,000 kilowatts. Electricity generation is inadequate to meet needs, a major cause of underutilization of capacity in Chinese manufacturing.

The government is well aware of the need to enhance fuel production and power generation in order to sustain a measured pace of expansion and to achieve the aim of doubling the country's GNP by the year 2000. To achieve this target, the country will need 1.4 billion tons of coal, 200 million tons of petroleum and 1,200 billion kilowatts of electricity by the turn of the century. China is determined to meet this challenge. A number of larger and medium-sized power stations will be constructed, among them more hydroelectric stations on the Yellow, Yangtze and Hongshin rivers. Thermal power stations will be located in coastal areas and along coal-transporting railroads. Nuclear power stations are planned to be located in areas short of other sources of energy. Large coal mines will open up in Inner Mongolia, Shianzi and Shanzi, while oil exploration will continue in Bohai Bay and other offshore areas and in the Tarmi Basin. The plan for the 1990s is to explore and develop new sources and to achieve efficiency in energy use.

Energy consumption per unit of gross value of industrial output fell at an average annual rate of 7 per cent in 1979-1981, 3 per cent in 1982-1983 and 7 per cent in 1984. Around 40 per cent of these energy savings stemmed from technical progress. However, about 60 per cent of energy savings was attributed to structural changes in terms of a decline in the importance of heavy industry and the closure of inefficient small-scale plants. It is important for China to take advantage of new devices which have been developed in other countries in order to enhance efficiency in the use of energy and material.

China's rural industry is emerging as a force in the industrial transformation. The ideological foundations of the rural economy of China have changed from central planning and large agricultural collective firms to market mechanisms and independent farms and firms. As a result, the rural industry has been the most rapidly growing segment of the Chinese industrial sector, filling numerous vacuums in existing demand. Rural enterprises also constitute an outlet for the surplus labour force. In 1989, 93.67 million persons representing 23 per cent of the total rural labour force worked in rural enterprises or 62 per cent of the increased labour force between 1978 and 1989. It was estimated that by 1990 rural enterprises would employ 100 million workers.

After a decade of rapid growth, rural industries currently face increasingly severe competition and market saturation, leading to a number of investment failures, inefficient technology and low quality of products, hindering the ability of rural enterprises to compete. Factor immobility instills a source of rigidity, inhibiting the diffusion and adoption of advanced technologies and practices. The inadequate network of support services is one of the constraints inhibiting the further growth even of successful enterprises. The average size of rural firms has not increased over the years. Although economies of scale do not necessarily constitute the determinant of efficiency, in the production of products such as cement and steel small size is generally a barrier to efficiency gains. As to the technological constraints, the problems differ between industrially developed regions and the least developed regions. In the developed regions the existing technological base is stronger. In the industrially less developed regions, there is a need to augment human skills to choose, adapt and to manage technology. The problem is not related to acquiring technology, but to choosing and adapting appropriate technology that befits local conditions. The implementation of the Spark Plan in the country's rural areas has promoted a number of local resource-based key industries. The "Spark Plan" is aimed at the gradual transformation of traditional agricultural patterns of scattered management into an intensive and operational type of farming, poultry farming, forestry and fruit growing in order to form commodity production centres. The Spark Plan particularly emphasizes the establishment of rural industries based on agricultural and mineral resources. The purpose is to combine crop production and livestock breeding with the processing of agricultural and animal products, and mining with the processing of mineral products. The Plan has injected new technologies into the production process of rural enterprises, helping them raise the quality of their products, reduce energy consumption and enhance the rational use of raw materials. Attention is also focused on preventing pollution in order to protect local ecology.

Between 1986 and 1989, over 20,000 projects were launched throughout the country within the framework of the Spark Plan, of which 10,346 projects have been completed. The resultant increased output value was to the tune of Rmb 22.18 billion. Exports from these projects fetched around $740 million in 1989, equivalent to 45 per cent of the total export volume of township enterprises in the preceding three years. Thus, the implementation of the Spark Plan in rural China ignites a fresh stimulus with advanced technologies yielding quick results. Given the massive government support for rural industries to evolve and adapt to the changing trends, the long-term prospects for rural industries are bright.

The largest single category of foreign direct investment hitherto has been co-operative joint ventures, accounting for almost half of the contracted value. Guangdong continues to be by far the most popular site, and investors from Hong Kong and Macao are the most active. Guangdong's proximity to Hong Kong's finance and services and the ease of monitoring investment just across the border are major reasons. Besides Hong Kong and Macao investors, only firms from the United States and Japan have contracted to invest more than $1 billion. None of the other investors exceeds $500 million. Investor interest in full ownership operations tends to increase and now accounts for as much as 6 per cent of direct investment. Most foreign funded projects were located in the Special Economic Zones (SEZs) - Shenzhen alone accounted for over half of contracts involving foreign investment. Since 1986, foreign investment has also tended to grow rapidly in the coastal cities, but the share of SEZs continues to remain dominant.

The regional dimension of industrial development in China is a pressing issue, given the uneven geographic distribution of the pace and pattern of industrialization. Until 1978, Chinese authorities placed great emphasis on the achievement of provincial self-reliance. This led to a relatively slow growth of factor mobility, a monopolization of regional markets by provincial enterprises, production duplication and a failure to fully exploit economies of scale. During the reform decade (1978-1988) China's industrialization strategy has changed from interior-oriented investments to treating the coastal region as a catalyst for industrial modernization. The future direction of regional development policy is now the subject of debate within China, a debate which invokes many of the key controversies facing planners.

The long-term objective of industrial policy in China is to put productivity growth on a sustainable path, to replace outmoded technologies and lower the incremental capital-output ratio in general, concentrating the effort initially on the key producers of major capital and consumer goods. The complexity of investment decisions, often involving piecemeal additions of modern equipment to old plants, generally lowers the returns from investment. The stress is currently on a gradualist, non-radical approach to industrial reform, on consolidation of changes already introduced and on an industrial policy that avoids putting severe pressure on the price level. Price reform is to proceed cautiously, but price control, implicitly by administrative means, will not be abandoned while scarcities persist.

The ownership structure of that part of industry currently in State hands will be maintained and efforts to resolve the contradiction between State ownership and incentive-driven management will focus on attempts to separate ownership from operational management. No large-scale privatization of the kind mooted by more radical reformers and placed at the centre of the economic agenda in eastern Europe are foreseen. The output targets set by the Eighth Plan (1991-1995) and for the ten years to the end of the century are generally modest by the standards of the reform decade. They are likely to be met and in many cases surpassed, provided that resources can be mobilized effectively. The current thrust of Chinese macroeconomic policy, with its stress on the need to slow the pace of growth in output and control demand, is cautious and conservative in tone; although reform is still identified as the goal of policy.

There is an increased awareness of the importance of sustainable development in the 1990s. The elimination of critical constraints in the energy, transportation, raw materials, basic industry and infrastructural sectors is crucial for sustaining an average annual industrial growth rate of 6.5 per cent enviaged for the period 1991-1995. Twenty programmes for technical co-operation have been designed within the framework of the Second Country Programme projects (1991-1995) and $49 million committed for the year 1991. The technical co-operation inputs are being directed to crucial areas of sustainable industrial growth.

I

THE ECONOMY OF CHINA

A. RECENT ECONOMIC TRENDS

The engine of economic growth in China is cooling down after showing visible signs of overheating in 1988. The pace of economic expansion eased to 3.9 per cent in 1989 and 5.0 per cent in 1990 (see Table I.1) after growing at or close to double-digit rates during 1986-1988. The first three years of the Seventh Plan (1986-1990) sustained the momentum of rapid growth which the Chinese economy has recorded since the introduction of structural reforms in the early 1980s. However, by 1988 there was serious overheating of the economy, exacerbated by the over-rapid expansion of industrial output and sharply rising consumer and investment demand fueling an outburst of inflation. Rising inflationary pressure, partly the result of price decontrol, which had appeared in 1984-1985 as well, was exacerbated by substantial wage and bonus awards to workers by enterprises.

The supply side of the economy was driven largely by the rapid growth of industry, while a 14.4 per cent average annual increase in investment in real terms had been the principal propellant of demand during 1980-1988. Backed by a 9.4 per cent and 7.4 per cent annual increase in government and private consumption, respectively, the surge in investment proved most dynamic despite repeated attempts to slow investment demand and curb planned expenditures by enterprises. Although the overall growth targets set by the Seventh Plan were surpassed, some macroeconomic imbalances surfaced during 1988-1990.

The government responded with policies to contain the growth of demand and to curb imports. The goal of economic structural reform, particularly the reform of a distorted price structure and the management of the key State-owned industrial sector, was of immediate concern in the interest of restoring macroeconomic stability. By means of a tight credit squeeze including higher interest rates, the wholesale cancellation of investment projects, and stringent controls on imports and on the allocation of scarce industrial inputs, measured consumer price inflation was brought down from close to 19 per cent in 1988-1989 to only 2.1 per cent in 1990 (see Annex Table A-1). Real growth of the money supply faltered to 1.3 per cent in 1990. However, efforts to hold back wage and bonus rises were generally less successful.

The external sector of the economy was also giving cause for concern. After a record trade deficit of $14.9 billion in 1985, export drives and more strict control over imports had pulled the current account back into near equilibrium in 1987, from which date imports started to soar again, resulting in sizeable current account deficits the following two years (see Annex Table A-2). To cut back the trade and current account deficits at a time of increased debt service obligations, the currency was twice devalued against the dollar, from Rmb 3.73 to Rmb 4.72 in December 1989 and again to Rmb 5.3 a year later. This and import controls helped to bring about a dramatic turnaround on the trade balance in 1990, when exports soared as domestic demand was squeezed and imports fell by nearly 10 per cent in dollar value. Good harvests in 1989 and especially 1990 (when grain output reached a record 435 million tons) also helped restore price stability, while the curb on investment and the very slow growth of industrial output during the second half of 1989 and the first quarter of 1990 to some extent eased pressure on scarce resources infrastructure - energy, transport and raw materials.

Table I.1. Key indicators of recent economic trends, 1988-1991
(Annual percentage change)

	1988	1989	1990	1991[a/]
GNP	10.8	3.9	5.0	5.5
Industry	20.7	8.5	7.6	8.6
Agriculture	3.9	3.1	7.6	3.5
Investment	23.3	-8.0	7.6	12.0
Export (FOB)	19.8	10.5	18.3	4.5
Import (CIF)	28.0	6.9	-9.6	11.2
Money supply (M_1)	20.3	18.0	27.8	20.8
Retail sales	27.8	8.9	2.5	10.1
Consumption	8.2	0.8	3.8	12.1
Inflation	18.5	17.8	2.1	6.2

Source: State Statistical Bureau of China, *China Statistical Yearbook* (various issues).
a/ Official forecast.

So sharply and successfully were the brakes applied to the Chinese economy during 1989-1990, that the authorities began to fear a full scale recession, with adverse effects on employment (millions of workers in the non-State sector and State sector alike were reported to have been laid off or be producing nothing in the wake of the austerity drive). By the fourth quarter of 1989 the government had decided to stimulate demand once more by reducing the high interest rates, and easing tight credit and strict investment controls. Industrial output grew at a modest (by Chinese standards) 7.6 per cent in real terms in 1990. Combined with a 7.6 per cent growth in agriculture, the highest during the Plan period, gross national product (GNP) was up by 5.0 per cent.

The restoration of some growth momentum after the stagnation induced by the austerity policies came at a price. Thanks to the need to inject working capital into many loss-making State-owned industrial enterprises, credit expansion accelerated to 22.5 per cent in 1990, faster than either of the previous two years. Even more serious was the further deterioration in the profitability, productivity and efficiency of State-owned industry in general and the 63 per cent overshoot in the budget deficit, as a result of falling tax and profit remittances, rising subsidies to industry to pay wages and cover loan payments and higher than expected costs in several other categories of expenditure. Because the savings rate remained high (and apparently increased) during 1990, the relaxation of austerity policies did not unleash domestic demand and consumer demand, in particular, remained weak as did investment, since a large proportion of the increased credit to industry was to fund working capital. This led to stockpiling of unsaleable goods on an unprecedented scale.

As China enters its Eighth Plan (1991-1995), macroeconomic stability has been restored and most importantly, inflationary pressures have been brought under control without sacrificing growth. A 5.5 per cent increase of GNP is forecast for 1991. The external balance is healthy and the decline in the growth of agriculture has been reversed. However, the threat of renewed inflation remains. Industry, especially the State-owned sector, is becoming less efficient and as much as a third of it is unprofitable, and the State budget is in chronic deficit. Population growth threatens to remain well above the rate which the economy can safely absorb[1/], and pressure on the infrastructure and on inputs to industry remains acute.

MACROECONOMIC TRENDS

Fig.I.1. Growth of gross national product (GNP) and industrial output, 1985 - 1990 (Percentage)

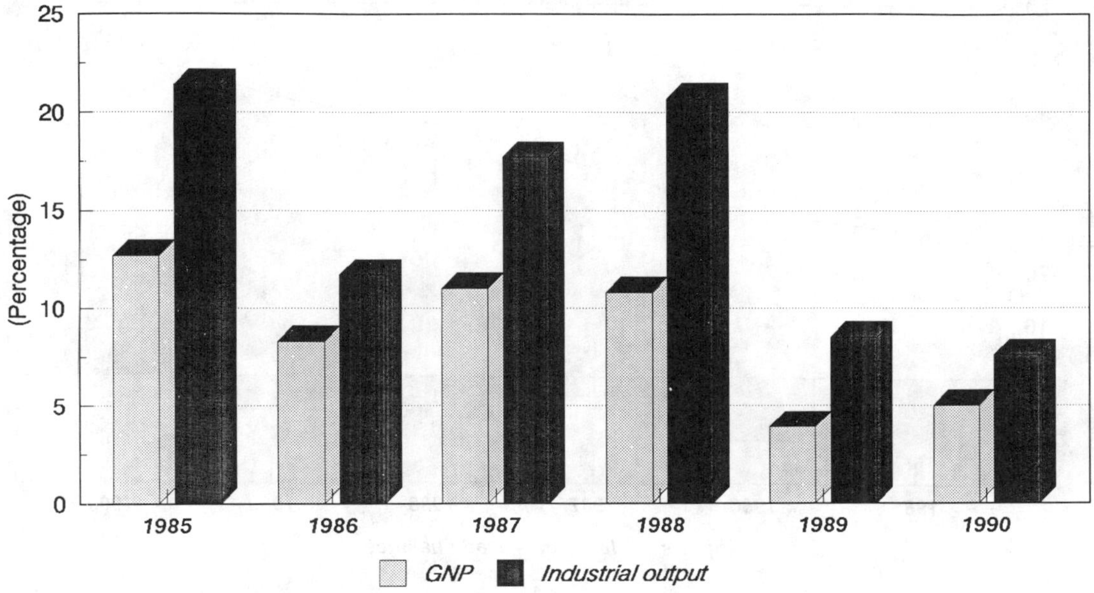

Fig.I.2. Structure of GNP by sector, 1970 - 1990, selected years (Percentage)

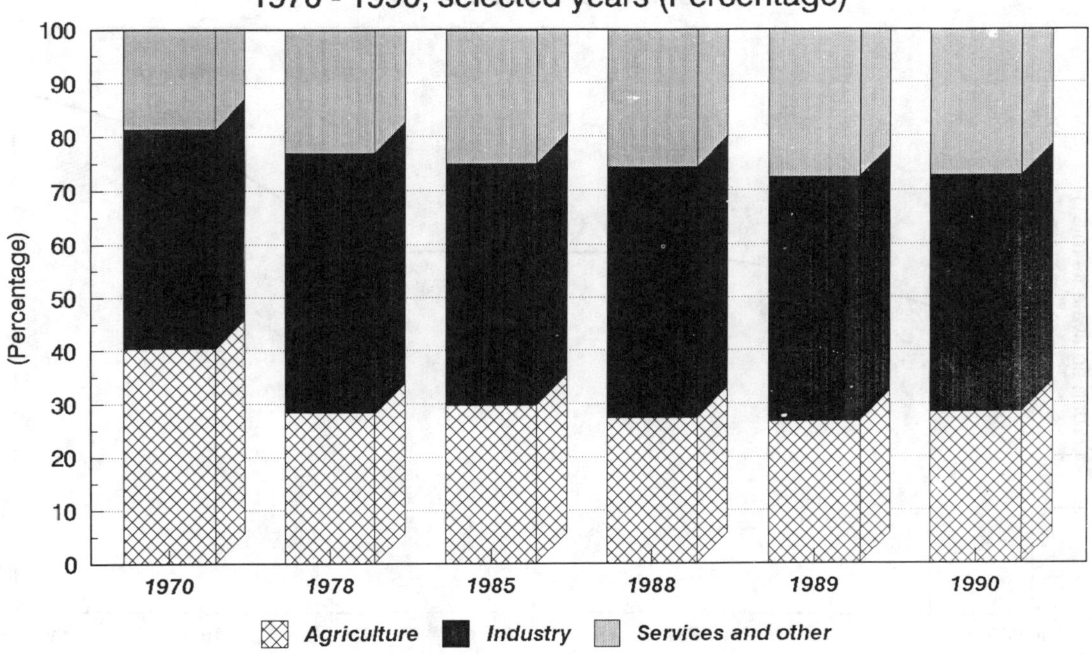

MACROECONOMIC TRENDS

Fig.I.3. Exports, imports and trade balance, 1985 - 1990 ($ billion)

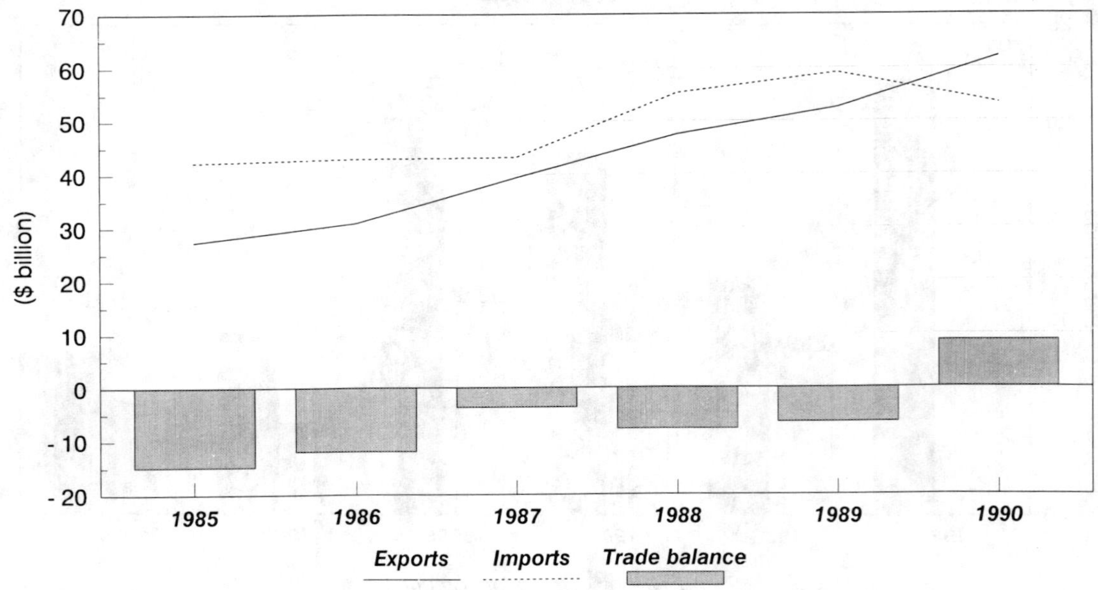

Fig.I.4. Exchange rate, 1985 - July 1991 (Yuan equivalents to $1)

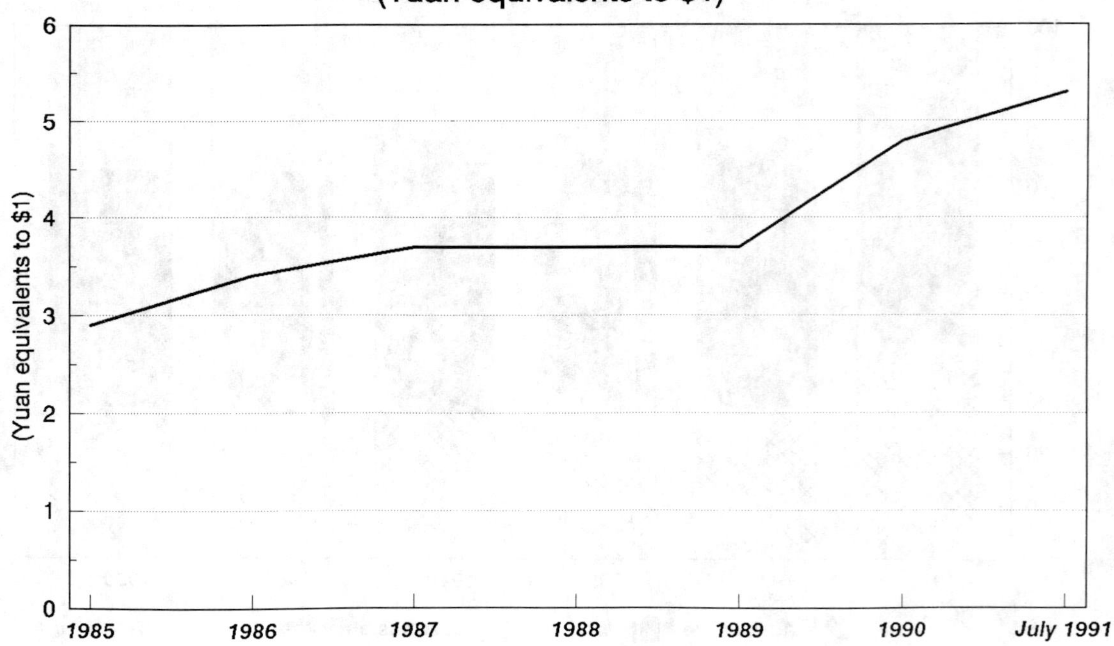

MACROECONOMIC TRENDS

Fig.I.5. External debt and debt service, 1985 - 1990 ($ billion and percentage)

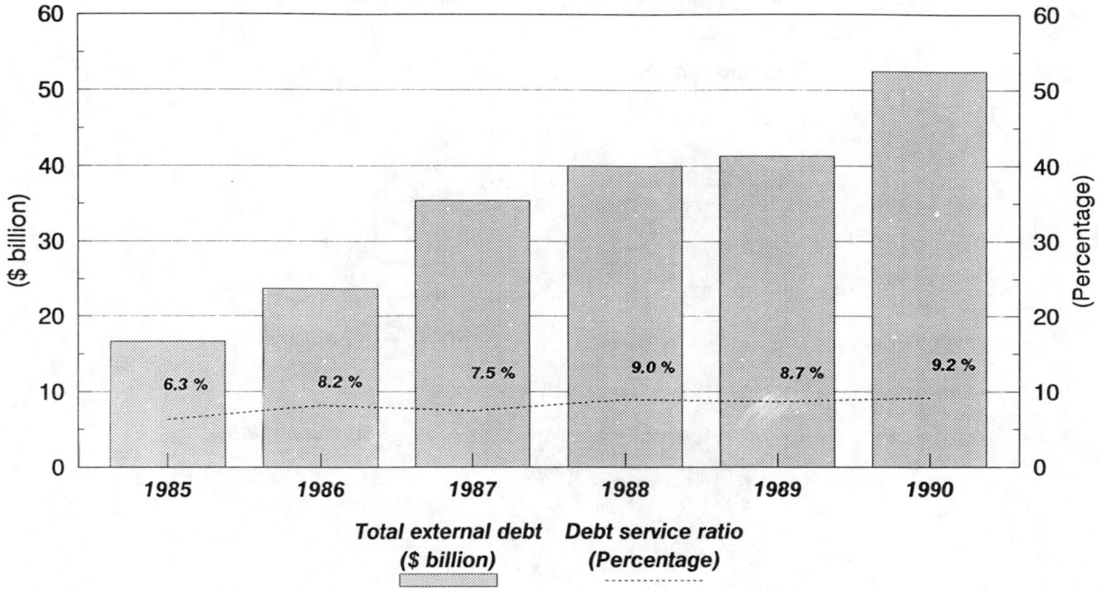

Fig.I.6. Rate of inflation, 1985 - 1990 (Percentage)

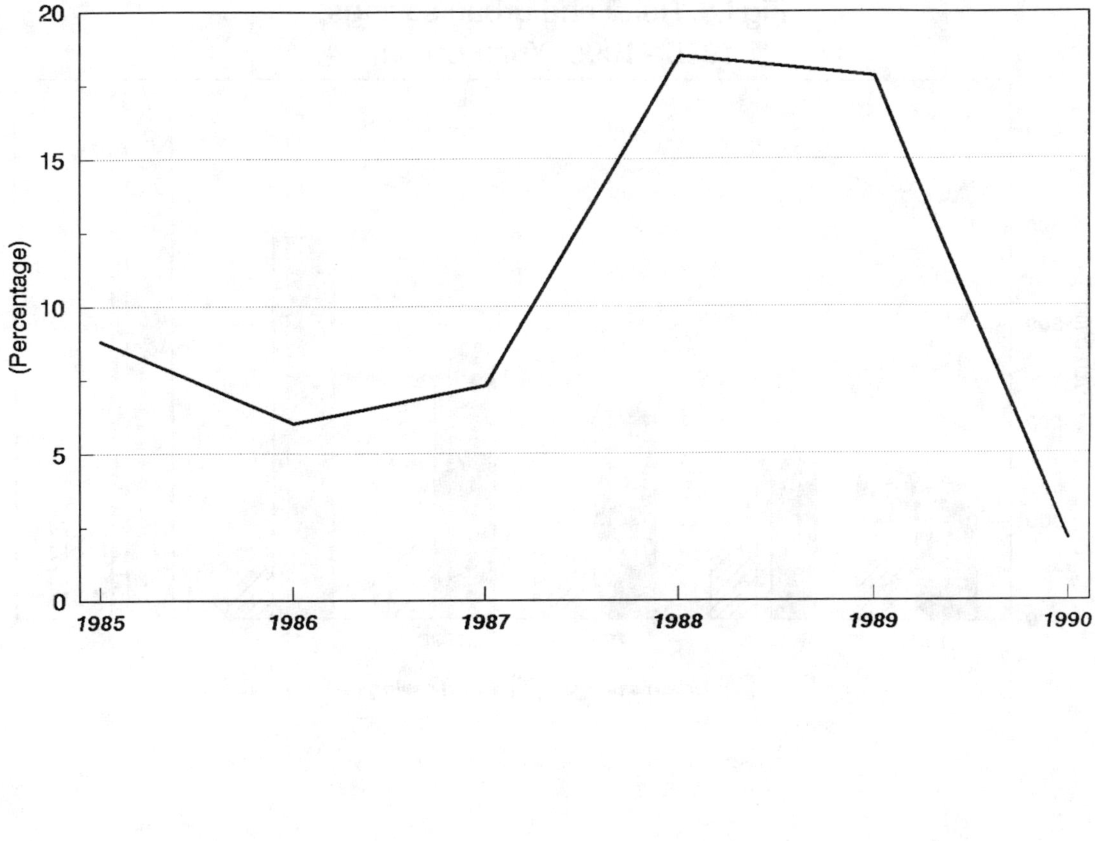

MACROECONOMIC TRENDS

Fig.I.7. Structure of employment by sector, 1990

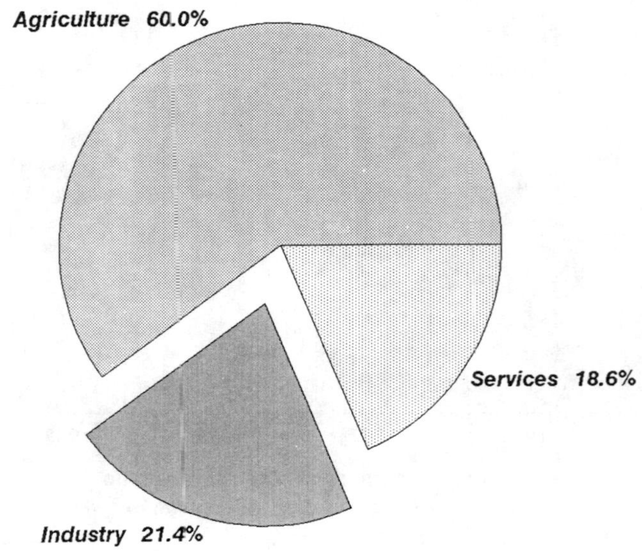

Fig.I.8. Rural and urban savings, 1985 - 1990 (Yuan billion)

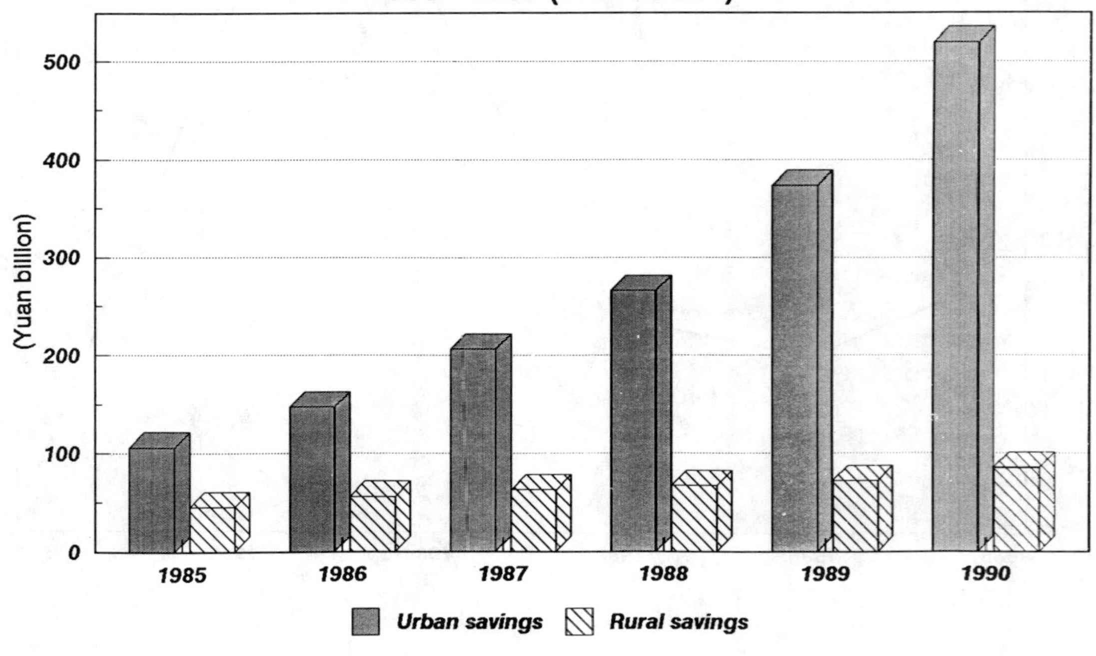

B. ECONOMIC STRUCTURE

Despite the rapid rate of economic growth during most of the 1980s China remains among the low-income economies on World Bank classifications. The World Bank estimates GNP per head at $330 in 1988, a figure which would have risen to about $350 in 1990 if the population had remained static and there had been no change in the exchange rate. Its long-term growth potential, though considerable in a favourable policy environment, is constrained by the demands on employment creation of a still rapidly growing labour force - rising by about 11 million a year - by severe skill shortages stemming from years of neglect of the education system especially during the Cultural Revolution (1966-1974), and by scarce or sub-optimally distributed natural resources. At about 0.1 sq km per head, cultivable land is scarce and becoming scarcer.

Agriculture

Agriculture is of vital importance to the Chinese economy, as a source of employment, to feed a population of over 1 billion, and to provide inputs for industry. The sector remains very important although its share of GNP has fallen over the years (see Table I.2). Despite the much higher growth rate of industry, 60 per cent of the labour force was defined as agricultural in 1988 (of which 78.5 per cent was actually engaged in agriculture). Of these, many have ceased to be full-time farmers, although some have returned to the land following the austerity measures introduced in the autumn of 1988 to curb demand. But the rapid growth of often very small-scale, privately or collectively run industrial undertakings - known as township and village enterprises (TVEs) - is a key factor behind the rapid annual expansion in real gross industrial output value (GIOV) between 1985 and 1989.[2/]

In keeping with its large size (roughly the same area as the United States), China's agriculture varies widely by area. Only a small proportion, about 10 per cent, of the total area is cultivated and about 45 per cent of this is irrigated, reflecting many centuries of intensive use of limited cultivable land. In the south of the country, where rainfall is abundant, triple cropping is possible. In the middle area of the country a two-crop system of rice and wheat is in place and in the cold, dry north, there is a one-crop wheat system.

Following the reforms introduced in 1979, in which the cultivation of land was effectively decollectivized under the contract responsibility system, procurement prices for staple crops were increased substantially and farmers were permitted to market a significant proportion of their output at market prices. Agricultural output at first grew rapidly; annual real output expanded by 8 per cent a year in 1981-1985. Thereafter, the real growth rate of agricultural output fell back to a more normal 3 per cent or so until 1989 and it is generally believed that a further sustained growth path can only be achieved by increased investment. (Data for gross fixed investment in agriculture are not available.) Investment by the State in agriculture has not been growing on anything like the scale required to raise agricultural productivity significantly and it is arguable that the sizeable funds in individual hands in the form of savings will not be channelled into investment on a large scale without another major rise in procurement prices, which would have serious implications for the State budget if they could not be passed on to urban consumers. During 1989, the austerity measures taken to cool an overheated economy led to the closure of perhaps 1 million TVEs. The resulting return of many of their employees to the farm implies a fall in the growth of agricultural productivity.

Industry

One of the most striking features of the Chinese economic structure, is the increasing preponderance of industry in terms of its share in GNP, which, at around 46 per cent in 1989, measured in current prices, was significantly higher than the 39 per cent which industry had contributed to total output in 1965. The faltering rate of growth of industrial output in 1990 resulted in a marginal fall in the share of industry in GNP to 44.3 per cent in 1990.

China's size and the development priorities of the past 40 years have made it a formidable industrial producer. Industry, which includes the extraction of natural resources, is divided into heavy and light categories. Heavy industry, which includes mining and logging, processing of raw materials and manufacturing of capital goods, accounted for 56.6 per cent of net industrial output in 1990; light industry for 43.4 per cent. Over the 30 years before the inauguration of the reform period in 1979, development policy tended to favour investment in heavy industry as China strove to develop a heavy industrial base in isolation from the rest of the world. At various times the bias towards heavy industry was temporarily reversed, but it was powerfully reinforced both by the withdrawal of Soviet aid and technicians in 1960 and by the perceived need to develop industrial (and defence industry) capacity across widely dispersed areas.

The vast majority of industrial activity (70 per cent of net value added in 1990) was still concentrated in the State-owned sector, either controlled by Ministries in Beijing or run by provincial governments. Industrial enterprises are classified as large, medium or small primarily on the basis of annual production capacity. In 1988, of the total 8.1 million recorded enterprises, 99,100 were State-owned large- and medium-sized enterprises. State-owned enterprises employed 18 per cent of the labour force in the same year.

There are also thousands of much smaller-scale plants using locally produced machinery and run under local control, as well as the large numbers of collective and private industries which have sprung up since 1979. These have traditionally been encouraged as a means of absorbing the growing rural labour force and have multiplied as a result of the encouragement of more diverse ownership and management structures during the reform decade. The number of enterprises owned by rural or urban collectives, or by individual families or a combination of the two, and the number of investment projects involving foreign investment have greatly increased.

The efficiency of Chinese industry in general has been impaired by a variety of structural factors. These include the ownership structure, the system of production according to quotas set by the State, the lack of financial and managerial autonomy, a pricing system which fails to reflect relative scarcities, a shortage of key inputs, overmanning on a large scale in accordance with socialist values, a poorly developed internal market and a shortage of skills. Evidence from analytical studies of the Chinese economy suggests that the incremental capital-output ratio in China is both higher than in other economies at a similar stage of development and rising, despite a considerable improvement in the quality of industrial plant over the last ten years.

Mining

China is adequately endowed with mineral resources. The problem with its reserves of oil and coal lies more with their location than with their quantity. The fact that industry, concentrated increasingly on the eastern seaboard with the abandonment of the third front policy during the 1980s, is far away from the mineral reserves found in abundance in the north and increasingly in the far west, poses strains on the overburdened railway network. The transportation network as a whole remains very underdeveloped and this further distorts cost/price relationships in the economy. Coal is very plentiful, providing over 70 per cent of primary energy. Onshore and offshore oil reserves are large, and more are being found. But output of both coal and oil has been lagging behind target and energy shortages, exacerbated by wasteful use, force many industries to operate well below capacity. Other mineral resources are abundant and China is currently a significant exporter of tungsten, antimony, tin and mercury and could enter the export market for other minerals. Imports of copper, aluminium and steel are inevitable as its reserves of iron ore are of low quality.

Services

Services account for a relatively low proportion of GNP - 27.3 per cent in 1990, in current prices. Nor has the share of services, which was estimated at 17 per cent in 1965, grown significantly. There is no doubt that the encouragement of economic activity outside the scope of the command economy since the early 1980s has increased the importance of services, and one of the reasons for their low measured share in output arises from the nature of the Chinese system of measuring economic output.[3/]

Table I.2. Output and employment structures, 1960-1990, selected years
(Percentage)

	1960	1970	1985	1988	1989	1990
Shares in GNP (In current prices)						
Agriculture	..	40.4	29.7	27.3	26.6	28.4
Industry	..	40.9	45.2	47.0	45.7	44.3
Other	..	18.7	25.1	25.7	27.7	27.3
Shares in labour force						
Agriculture	62.2	80.8	62.5	61.1	60.2	60.0
Industry	20.6	10.2	21.1	22.1	21.9	21.4
Services	17.2	9.0	16.4	16.8	17.9	18.6

Source: State Statistical Bureau of China.

Finance and fiscal structure

The role of the banking sector in managing the economy has expanded in recent years, although the financial system remains fairly rudimentary, and interest rates are adjusted infrequently and were not used as a form of macroeconomic policy until the austerity drive of 1988-1989. The spread between what is paid to depositors and charged to borrowers is often zero and decisions on the granting of credit to the industrial sector are frequently made on the basis of non-economic criteria.

Despite this, banks and rural credit cooperatives are playing an important role, with the Industrial and Commercial Bank, for example, now having the management of commercial and industrial credits and deposits from the public, and specialized banks engaging in lending for construction projects and to the agricultural sector.

The rapid expansion of the Chinese economy in recent years has been accompanied by speedy growth of the money supply and of deposits (see Annex Table A-3). Growth of the latter has accelerated particularly markedly in 1989-1990 as people saved both to restore the real value of their cash balances in the face of inflation and also perhaps because of a lack of goods on which to spend as production of the most desired consumer goods stagnated and much sought-after imports became scarce and more expensive under the austerity policies. An element of forced saving is also occurring through the issuance of government bonds to finance the burgeoning deficits being incurred by the Central Government. These are distributed to the provinces on a quota basis and individual enterprises then "sell" them on to workers.

A domestic savings rate of 38 per cent of GDP was estimated for 1987,[4/] half of which came from household savings and half from enterprises. Among the factors prompting rapid growth in household savings during the reform decade have been the dismantling of the commune system and the removal of some of the social security functions provided by the workplace before the reforms, rapidly rising incomes and a scarcity of consumer goods on which to spend accumulated cash balances. The freedom to own property, especially housing, in the countryside, has been a powerful stimulus to rural savings, as families defer consumption in order to build houses; this has been reinforced by a secular trend for the urban-rural terms of trade to move in favour of the rural population, as procurement prices for agricultural staples have risen and farmers have been able to market an increasing proportion of their output at market prices. The rural savings rate was estimated at 26 per cent in 1987, of which nearly half was destined for housing. The urban savings rate, which was only 3.5 per cent in the same year, is much lower partly because there is as yet no widespread property market in the cities. The urban population also has privileged access to a number of State-provided services through the role of the workplace and spends a larger proportion of its disposable income on food.

The devolution of many decision-making, resource allocation and revenue collecting powers to the provinces and lower levels of government, which has proceeded apace since the beginning of the reform decade, has left the Central Government's resources severely strained. The revenue base has narrowed the majority of enterprises now paying tax, rather than handing over all their profits to the State, since the yield from such taxation is limited by a process of negotiation which fixes the amount paid in tax in cash terms under the so-called contract management responsibility system. There are a number of different taxes, levied on individuals as well as enterprises (with the latter being the preponderant source of revenue and accounting for about 40 per cent of total tax revenues), including taxes on profits, turnover and property. Many of the taxes are collected by local governments.[57]

The fiscal system aggregates all the revenue and spending of each level of administration and the majority of the Central Government's income consists of subventions from individual provinces, the amounts negotiated in a complex procedure. The Central Government's direct revenues are less buoyant than those of the surplus provinces partly because the industrial concerns to whose taxes it has access tend to be in the less profitable, heavy industrial sector, which is still subject to price control. A complicated array of revenue-sharing arrangements exists among the different provinces, some apparently handing over a fixed sum in money terms, others a proportion of total revenues. The more wealthy provinces have so far apparently been successful in resisting changes to this *ad hoc* system, although the Central Government has recently been able to assert greater control over imports and to increase its share of foreign exchange earnings from exports. The Central Government, which has run a deficit since 1979, gives annual grants to the poor, deficit provinces in the underdeveloped interior regions as well as normal grants to all provinces for specific purposes. It also controls defence expenditure and very large capital projects.

The share of the State in GNP has decreased over the period of the Seventh Five-Year Plan, while the budget deficit has soared. The burden placed by subsidies (Rmb 57.9 billion for enterprise losses and Rmb 37.9 billion to compensate for price rises in 1990, or nearly one third of estimated outlays of Rmb 357.2 billion in that year) is particularly acute, and servicing the public debt is becoming a burden. The government borrows domestically and abroad to fund the deficit, which soared to Rmb 15 billion in 1990. Efforts to curb capital expenditure have resulted in the share of this item in total spending falling to 19 per cent in 1990, from 32 per cent in 1985. In the short term, this is not a concern, reflecting as it does some success in controlling the construction and investment spending of local governments. Efforts to limit the growth of the deficit, to reassert a greater degree of central control over local spending and revenue collection, and to widen the very narrow tax base, are regarded as priority goals by the Central Government. China's foreign debt has grown considerably, rising from some $16.7 billion to over $52 billion between 1985 and 1990, with the debt-service ratio rising from 6.3 per cent in 1985 to 9.2 per cent in 1990.

Consumption, investment and trade

China's economic structure reflects the priorities of the command economy and the perceived need to limit consumption in favour of investment over a long period. Consumption, at 63 per cent of 1988 GDP, takes a lower share of demand than India's 79 per cent. Investment and savings are both high; in 1988 investment exceeded savings, contributing to the current account deficit which was a feature of the external accounts until 1990. Reflecting China's huge size and known self-reliant development priorities since 1949, its economy remained relatively closed. In 1988 exports accounted for only 7 per cent of demand. This share has been growing rapidly in recent years, as China's links with the world economy have become closer. The value of trade (exports and imports) accounted for roughly 30.6 per cent of the national income in 1990. The share of exports in 1990 at $60.1 billion represented around 16 per cent of GDP, while that of imports ($53.4

billion in 1990) was 14.6 per cent. A significant rise in imports shows that China has become an important export market for many countries.

Although, apart from the export-oriented coastal regions, the Chinese economy as a whole has a relatively low dependence on trade, the value of trade turnover has increased very rapidly since the adoption of the "Open Door" policies in the early 1980s. Attempts have been made to ensure that imports and exports are roughly in balance, but periodic relaxations of control on imports have led to soaring demand for both consumer and capital goods imports from abroad, resulting in trade deficits from 1985 until 1990. Ever since 1985, indeed, efforts have been made to curb imports, by devaluation and by strict controls, but it was not until domestic demand was severely curtailed in 1989-1990 that the trade account returned to surplus. The clampdown on imports, which are needed as inputs to various manufacturing processes, does not seem compatible with the planned level of economic growth. Running a trade deficit is perhaps inevitable in order to sustain even the measured pace of economic expansion and to fulfill the demands of the modernization drive.

For most of the 1980s China sacrificed some growth by allowing petroleum (mainly crude oil exported to Japan) a commodity in short supply - to remain the single largest foreign exchange earner, and was penalized by the low prices obtainable for oil during that period. From 25 per cent of export revenues in 1985, oil fell to only 6.8 per cent in 1989. More than half of China's exports are now manufactures (generally of relatively low value added) and the majority of these are of clothing and textiles, a typical developing country's export profile. Other exports include agricultural products (notably foodstuffs to the Territory of Hong Kong and Japan), raw materials and coal.

Grain imports needed (imports of which have ranged from a low of 6 million tons in 1985 to a high of 16.6 million tons in 1989) to feed the populous eastern cities and the requirements of the modernization effort dominate the import mix. Iron and steel (which China produces in large quantities, but not sufficiently in the desired quality), machinery, electronics and transport equipment are the main imports. Imports of consumer goods have been stringently controlled since 1985.

In 1987, Hong Kong overtook Japan as China's biggest trading partner and is the largest market for Chinese exports and the largest single source of foreign exchange. A significant share of exports to (and imports from) Hong Kong is destined for or originates in third countries, reflecting both China's role as a low cost manufacturing base for Hong Kong manufacturers and trade with countries with which China has not had official trading relations. The latter aspect of Hong Kong's entrepôt role may well diminish as the number of countries with which China has direct trading links increases. Other important trading partners apart from Japan are the United States and selected EC countries. Trade with the USSR, having stagnated since the 1960s, is now growing quite rapidly.

One of the major planks of the modernization effort has been the acquisition of modern technology by attracting direct foreign investment. China has indeed attracted significant, although not massive, foreign funds during the 1980s, with a total committed inflow of $78.5 billion against $47.7 billion actually utilized. The chief means of attracting foreign capital in the form of direct investment has been through the establishment of special zones offering various tax concessions and preferential access to scarce inputs. As part of the decentralization process, many provincial governments have sought to attract such investment by setting up similar incentives. But the intention of the State has been to concentrate such endeavours on the eastern seaboard, and after the establishment of the first four Special Economic Zones in the south-east in 1980, 14 coastal cities along the eastern seaboard were opened on similar terms and the island of Hainan was made into a province offering investment incentives a little later. At first foreign investment in China was concentrated on the oil sector, when, in 1982, 27 foreign companies signed contracts for offshore drilling rights. As well as oil and gas, foreign investment has been sought for other priority sectors, notably coal, nuclear energy and some large-scale industrial projects.

Some large and well-publicized projects have been completed and the Chinese authorities have successively improved the terms offered to foreign investors. A key area of negotiation has been over access to China's potentially huge, but currently underdeveloped, domestic market. Foreign investors have been aiming to penetrate this market while the Chinese have export-oriented investment. Another issue has been a Chinese perception that the level of technological transfer, especially from Japanese companies, has been insufficient.

The major success of the foreign investment drive has been the attraction of large amounts of investment, often in individually modest amounts, from Hong Kong and Taiwan Province,[6/] to the zones and investment areas close to Hong Kong and across the strait from Taiwan Province. Hong Kong has accounted for the lion's share of direct investment, much of which has gone to Guangdong province to set up light industrial export manufacturing bases, often operating a low level of technology and attractive primarily because of low labour costs.

As well as seeking direct investment from abroad, China is also a significant recipient of credits from institutions such as the World Bank, and of bilateral lending programmes, notably by Japan. As discussed above, the foreign debt, although of considerable size at about $52 billion, is mostly on concessional terms and the debt service burden moderate. The invisibles and transfers headings of the current account have generally been in surplus thanks to travel, labour service and transfer surpluses, but there will be a peak debt repayment on medium-term finance, requiring outgoings of about $10 billion a year, which may give rise to current account deficits.

C. KEY ISSUES

Rectification and economic slowdown

The reform process envisaged by the Chinese leadership during the 1980s was based on pragmatic motivation, with the process of reform being seen as a means to aid in the modernization goals set forth as long ago as the 1970s rather than as an end in itself. The aim has been to reinvigorate the economy by selective combination of elements of the market with the planned economy. The reforms have been generally successful in unleashing economic dynamism in the rural sector, among the newly burgeoning collective and private, generally small-scale enterprises and especially in the open areas of the coastal region. As the process of reform gathered momentum, up to 1985, the question of how to invigorate the key State-owned sector and reform the urban economy became increasingly pressing. But the government has been hesitant, in view of the enormous difficulties, to act decisively on the important issue of management/ownership of the State-owned industrial sector and on fundamental reform to the price structure. Debates about these key issues have yet to be definitively resolved. Like other socialist countries which have attempted to mix the planned economy with elements of the market, China has found the process of economic reform problematic, the reforms themselves having thrown up various macroeconomic and microeconomic dilemmas.

China's economy can still be characterized as one of scarcity, in which demand pressures, once unleashed by reforms, drive prices upwards, a process that the plethora of price controls and the State allocative apparatus can temporarily restrain, but which can only be effectively articulated (though not brought to a halt) by demand management. As the preceding discussion shows, the leadership has so far only been willing partially to manage demand, and it has been the more dynamic collective and privately managed industrial sectors and the rural peasantry which have borne the brunt of the austerity measures.

In 1985-1986, after the emergence of a widening trade deficit, mounting urban inflation and a fall in grain output, pragmatists and reformers alike adopted a more cautious approach to economic management and postponed the enactment of radical policies designed fundamentally to change the economic management of enterprises. But many of the more radical supporters of the reform programme, especially academic economists, argued that the best way forward was to accelerate the pace of change, rather than to slow it. Nevertheless, the Seventh Five-Year Plan (1986-1990) called for up to two years of rectification and slower growth.

The first serious challenge to the policies and personalities associated with the reforms did not come until 1987. The period 1987-1988 marked the high tide of the influence of more radical reformers, despite the overheating already evident in the economy at that time. Political reform was placed on the agenda at the National People's Congress, which envisaged removing government and economic management from the day to day control of the Party and the establishment of a body of laws defining the rights of individuals *vis-à-vis* State and Party.

In the spring of 1988, control over the urban prices of pork, vegetables, eggs and sugar was lifted, sparking a wave of panic buying in the cities as people anticipated further price rises. As mentioned earlier, this added to inflationary pressures (see Annex Table A-4) fuelled by large wage and bonus increases in State-owned enterprises and the increasing shortage of key inputs to industry and led the authorities to decide on the postponement of further price reform, initially for six months. Further austerity measures swiftly followed in a retreat from reform that has yet to be reversed and was strengthened by the political traumas of the summer of 1989.

Official policy, while still espousing economic reform as a long-term goal, still stresses the need for rectification and retrenchment and the outlook for the reform policies is uncertain. As analyzed in a document released in January 1990 the Decision on Further Improving the Economic Environment, Straightening Out the Economic Order, and Deepening the Reforms, China's economic problems can be listed as follows:

- Demand running far in excess of supply;
- Real national income growing by 70 per cent between 1984 and 1988; capital investment by 214 per cent; cash incomes by 200 per cent. The growth was financed by deficit financing, expansion of the money supply, domestic credit and running down foreign exchange reserves;
- Agriculture failing to cope with the scale of industrial expansion;
- Primary industry and infrastructure failing to keep pace with the rapid development of processing industries, causing chronic bottle-necks in infrastructure and acute shortages of coal, power, oil and steel and forcing much of industry to lie idle; and
- Inordinate decentralization in the distribution of finance and materials.

The major initiatives proposed to deal with these problems, which still form the basis of current economic policy, were outlined as follows:

- Control of demand by means of strict regulation of investment, tight wages and bonuses and credit;[7/]
- Strengthening agriculture and basic industries. Investment in agriculture is to be increased. Energy output is to be raised and the price of coal increased. Priority allocation of key inputs in short supply is to go to selected State-owned enterprises under the double guarantee system which provides for guaranteed access to inputs in return for assured delivery of taxes to the Treasury under the contract management responsibility system;
- Two-tier pricing, under which many goods have a low, subsidized price if produced under the Plan and a higher market price for any surplus, is to be eliminated. Efforts to raise State prices gradually towards market levels and subject the new unified prices to ceilings appear to be envisaged in an attempt to prevent accelerating price inflation. This implies a major retreat from the goal of price reform; and
- Import controls, but with the proviso that the Special Economic Zones and open cities which have attracted the lion's share of foreign direct investment will be allowed a more liberal regime.

It was broadly agreed by the leadership in 1988 that the growth momentum of the previous few years was unsustainable and that a more measured pace of economic expansion must be pursued. But policymakers remained divided about the future of reforms. The divisions were reflected in various apparently contradictory moves made during 1989 and 1990. Some of these, such as an order that the government involvement in enterprises with foreign investment be strengthened, appeared conservative in tone, while others - allowing the embryonic stock market in Shanghai to gather under one roof, lifting price controls over more goods and services in early 1991 - have a reformist hue. Of course conservatives and reformers at the centre of government are united in wishing to see a reduction in the burden of subsidies on the State budget and there appears to be a consensus in Beijing favouring a measure of recentralization of key aspects of resource allocation.

The proposals, discussed at the meeting of the National People's Congress in March-April 1991, for the Eighth Five-Year Plan which has already commenced are the fullest currently available guide to policy. Reflecting as they appear to do a compromise, they are a broad outline of goals for the next five and ten years.

Growth is targeted at 6 per cent a year, for the Eighth Five-Year Plan and the Ten-Year Development Strategy (1991-2000), with the proviso, as in the Seventh Plan, that there will be a further period of rectification and austerity before growth momentum will be resumed.

Specific sectoral targets are modest. Although electricity output is slated to rise by 5.7 per cent a year in 1991-1995, the targets for coal (which still accounts for about 70 per cent of electricity) and crude oil are only 2.3 per cent and 1.0 per cent a year, respectively. This calls for significant improvemnts in energy efficiency.

With the proviso that fewer projects are to be started than during the previous five years, the priority sectors for investment are identified as agriculture, including water management, basic industries, energy, and infrastructure - not a markedly different emphasis from the previous Plan. The draft proposals themselves do not specify amounts, but an official comment has been made envisaging average investment worth Rmb 520 billion a year. This is more than the Rmb 398 billion a year of the Seventh Plan, but represents a fairly small real rise in view of the price inflation in investment goods, which has exceeded that for consumer goods.

On the questions of economic management and reform the proposals are not specific. The general emphasis is on deepening reform and perfecting the various systems already in place in State-run enterprises. While acknowledging a need to do something to cut the State's growing budget deficit, and advocating a return to surplus on the current account, the proposals say almost nothing about how the Central Government will increase its share of revenue.

The proposals commit China to the maintenance of the "Open Door" policy and also, in what appears to have been an acknowledgement of previous neglect, enshrine as a key project the development of the Pudong area of Shanghai as an open, technological zone.

Other proposals related to the question of wage and price reform are vague, but suggest continuity of policy, rather than the abandonment of aspects of price reform apparently envisaged in the January 1990 document.

On the subject of TVEs, the proposals are sketchy but the development of services is to be promoted, so that the tertiary sector's share in GNP rises from about a quarter now to a third in the year 2000. This partly explains why the growth rates for specific products are so low and is a theme which was taken up in the Seventh Five-Year Plan. One aspect of promoting the services sector, apart from resource considerations, is its employment creating potential.

Perhaps the most significant discontinuity with the Seventh Plan concerns population growth. The proposals stress the need to limit the natural growth of the population to 1.25 per cent a year, while the Seventh Plan was much more ambitious, aiming to hold population growth to 1.14 per cent a year. The overshoot during the Seventh Plan, when the population grew by 1.54 per cent a year, has very serious implications for policy, especially employment creation and land use. The Eighth Plan proposals tackle the subject of regionalism by the observation that the more developed areas on the eastern seaboard will concentrate on moving further up the value added scale, while the resource-rich provinces will seek to develop processing industries.

Regional disparities

China's vast size and the uneven dispersal of resources and population make regional disparities in terms of wealth and income inevitable, although the government has sought to mitigate these by redistributory policies and special aid to relatively backward areas as well as an egalitarian welfare system, particularly in State-owned enterprises.

Regional development policies in the pre-reform era reduced disparities between provinces by aid to agriculture in the less developed areas and by the creation of new industrial centres, with more regard for strategic and social concerns than for comparative advantage. A priority of the government was to link previously inaccessible provinces in the interior into the national rail network. After the withdrawal of Soviet aid in 1960, the desire to disperse industrial resources was powerfully reinforced and the Third Front policies dominated decisions about the location of new industries in the 1960s and 1970s. Areas not related to this programme, industry in the more developed east and agriculture in general, were neglected.

These policies produced a relatively even income distribution and a fairly impressive growth rate of 4 per cent in real national income per head over the long period 1952-1982. But the neglect of agriculture, especially in the less developed regions, the extremely low concentration ratios of key industries, the local (often province-wide) protection offered to industrial enterprises, the barriers to entry by new producers and the disproportionate cost of greenfield development held back the growth of productivity.

Thanks to the agricultural reforms of the early 1980s the peasantry in general benefited from increased procurement prices. However, those best placed to exploit the new opportunities to market their above quota-produce and to diversify out of grain production into "sideline activities" such as animal husbandry, aquaculture or horticulture, where the returns are higher than for grain, saw their real incomes rise much more sharply. Thus the disparity in income per head of the peasantry between remote and poorly endowed Gansu province and the fertile Yangtze valley near Shanghai widened from Rmb 98.4/Rmb 290 (2.9) in 1978, on the eve of the reforms to Rmb 340/Rmb 1,301 (3.8) in 1988, when *per capita* incomes rose by 13 per cent a year in nominal terms in Gansu and by 16 per cent in Shanghai. The new opportunities opened up by the reform era have also widened disparities within regions as well as between them, with the peasantry farming in the rural areas around large cities being particularly well placed.

The State has not resisted the emergence of these disparities. But the government has been quite careful both to protect the real incomes of the urban population, whose standard of living is much higher, and to prevent income disparities in the cities from becoming too wide. In 1988, the very wide variations in terms of gross industrial output per head were not fully reflected in average wages (including subsidies and bonuses). Premium wages are paid to workers in exceptionally poor and backward areas like Tibet, and the north-western provinces. The gap between the lowest and highest wages per head in 1988, in Guangdong (Rmb 2,618) and Shanxi (1,185), at a ratio of 2.2, was much narrower than in the case of farmers.

The care with which the relative equality of urban incomes has been maintained, at a huge cost to the State budget in subsidies, has helped to moderate the powerful centrifugal forces at work in the Chinese economy and polity, forces which have been strengthened during the reform period, in which each province and locality has striven to maximize the opportunities for local growth.

The transport network has remained underdeveloped in a country criss-crossed by natural barriers to mobility. In 1986, the rail density (in km per thousand of population and km per thousand square kilometres (sq km)) compared unfavourably with India (0.05 and 5.5 versus 0.08 and 18.9) as did the road density (0.9 and 100.7 versus 2.3 and 538.9). In addition, a rigid system of social control, recently somewhat relaxed, has prevented inter-provincial migration and strictly limited rural-urban migration, in search of work.

The administrative division of China, with its provincial and lower level government structures mirroring the Central Government, has also, by placing large bureaucratic structures at multiple levels, played a part in facilitating the evolution of self-reliant development policies by local administrations anxious to maximize the benefits to their constituents. There are indications of provinces and sub-provincial localities erecting barriers, many of them illegal, to internal trade and levying extra-legal taxes on local industry, in defiance of prohibitions from the Government in Beijing.

In the reform decade the bias towards investment in the interior provinces, at the expense of the more developed eastern seaboard, largely came to an end and the State allowed the pace of expansion of the coastal area to accelerate. During the middle to late 1980s, therefore, the north-eastern provinces including Hebei, and the municipalities of Beijing and Tianjin, saw a below average growth in investment. Investment in some of the poorer interior provinces - Guizhou, Yunnan, Gansu, Shaanxi and Inner Mongolia - also expanded at a rate below the national average. Major beneficiaries of new investment were in the south and south-east, in those areas whose location and human infrastructure made them best placed to take advantage of the new opportunities, especially in foreign trade and investment. These provinces also recorded above average growth in real industrial output, with Jiangsu, Fujian, Jiangxi, Zhejiang, Guangdong and Shandong recording annual average growth of over 15 per cent per year in the period 1983-1988.

The three major municipalities - Beijing, Tianjin and Shanghai - grew more slowly, partly due to constraints imposed by infrastructure. Shanghai has consistently been starved of investment funds under the post-1949 Government. Nevertheless, its *per capita* gross industrial output still exceeded the national average by a factor of 6 in 1988, despite the fact that its gross industrial output expanded by only 5.6 per cent a year in 1983-1988, about half the national average of 10.4 per cent. Shanghai has argued that its slower growth owes much to the relatively large share of tax revenues (about 70 per cent) that it must hand over to the central authorities. Its overstrained infrastructure has also served as a disincentive to foreign investment, despite its good human resource base (a far larger percentage of the workforce is highly educated in Shanghai than in any other Chinese city).

Apart from Shanghai, the rest of the east and south-east of China has experienced boom conditions during the 1980s, both in development of lucrative sideline agriculture and in terms of growth in industrial output. Much of the new industrial investment in this coastal area was in export-oriented light industry, a significant proportion of it in the form of manufacturing or processing arrangements with foreign buyers and investors, particularly from Hong Kong.

So vibrant indeed has been the export-oriented coastal economy, especially those areas closest to the entrepôt of Hong Kong, that this part of China is often viewed, economically, as belonging to the ranks of Newly Industrializing Economies (NIEs), forming part of a "greater China" that embraces Hong Kong, Taiwan Province and the coastal part of the mainland. As such its economic fortunes depend more on the world trade cycle than on developments in the rest of China, a state of affairs which is exacerbated by the weak linkages across the economy.

With economic buoyancy has come a greater degree of autonomy for this part of China, which has tended to resist the austerity policies advocated by the Government in Beijing and has been able, so far, to secure a significant part of the revenues from trade to itself.

D. ECONOMIC PROSPECTS

In view of the relaxation of controls over credit in effect during 1990 the growth of industrial output is likely to remain fairly robust during 1991, especially in the private and foreign-funded sectors (which only accounted for 8 per cent of total industrial output in 1989). In the first five months of 1991, industrial output rose by 13.3 per cent with rural industries growing at 31.6 per cent. The output of State industries rose only by 9.2 per cent. Collective enterprises recorded a 19.0 per cent increase in output during the same period. Amidst rising industrial output industrial inefficiency continued to persist and caused a 34.0 per cent drop in real profits between January and April 1991. Since the government, in line with austerity, has announced a target of 4.5 per cent for GNP growth in 1991, and in view of the rapid expansion of credit during 1990 and early 1991 a further squeeze on industry, especially outside the State sector, can be expected during the next 12 months in order to curb inflation. This will probably be similar in scope to the austerity policies nominally in force since 1988 and the most important "backbone" State enterprises will be spared cutbacks in production and labour via the "double guarantee system". Meanwhile attempts to crank up productivity growth in State-owned enterprises will have to focus on investment and on the control of excess wage and bonus payments.

An overall growth rate of industry of about 6 per cent in the first two years of the current Plan would be consistent with holding GNP growth in the 4-5 per cent range, before allowing expansion to gather momentum. In the long term, industrial output has the potential to grow very much more rapidly, but its medium-term growth rate is constrained by infrastructural bottlenecks and shortages of inputs. The former will take many years to correct and the latter can be partially alleviated by allowing imports to resume growth, subject to the further constraint that the current account deficit does not become burdensome. For the next ten years therefore, it seems likely that attempts will be made to keep the growth of gross industrial output below 10 per cent a year, with the pattern skewed by location and ownership. The dynamic export-processing regions will continue to be buoyant, with their growth more a function of world trade than of domestic policy. The fortunes of TVEs will fluctuate in accordance with the general macroeconomic environment. They will be periodically subject to restraints via curbs on lending growth and on access to raw materials. The outlook for State-owned industry is problematic; its financial health will need considerable amelioration either from the State budget, or (less probably) from significant price reforms before it can be allowed to resume growth patterns typical of the 1980s.

Agriculture, which grew at an unsustainable 7.6 per cent in 1990, is believed capable of a trend growth rate of about 3.5 per cent and the government is likely to do all in its power to sustain this, both to feed the growing population and to provide inputs for processing industry.

As noted above, the employment-creating potential of the service sector has led the government to call for tertiary industry to expand more rapidly than agriculture or industry. The means of achieving this are unclear, but it certainly implies continued encouragement of individual forms of ownership and will depend on sustained growth in consumer spending power, suggesting that the high savings rates of the 1980s may have to decline somewhat. In view of the very subdued level of consumer demand during 1989-1990, during which time savings increased, a fall in the next few years would not be unwelcome, but will depend on more effective production and distribution of domestic output or relaxation of controls on imports.

Only a moderately more favourable climate for imports can be expected to prevail in the next few years, as the State remains determined to conserve foreign exchange to serve the needs of the modernization programme. Export growth, however, should continue to be buoyant provided that China does not fall foul of protectionist moves in the west and world trade growth does not falter. The coastal regions will remain at the forefront of the export drive, which will continue to be most successful in light industrial consumer products which are highly price-elastic on world markets.

The likely evolution of policy, especially towards resuming the momentum of reform, is particularly difficult to predict in the current climate. On the one hand, it appears evident that the reforms so far introduced cannot be abandoned. On the other hand it is hard to discern a strong commitment to reform among the leadership, particularly as regards the sensitive areas of price and State-owned enterprises. Further ahead it seems probable that some momentum will be regained, if for no other reason than as the only way to enhance the productivity and efficiency of industry. But changes are likely to be gradual and hesitant and may not be ventured on any scale during the next few years.

An immediate and pressing concern, shared by many, is the state of government, especially Central Government, finances. Measures to deal with this will be politically complex, but it is probable that there will be significant changes to the current revenue-sharing arrangements, which will themselves tend to favour reforms, as the role of the financial sector and the tax system will be key determinants in balancing the State's finances.

A major area of uncertainty is the evolution of regional policy but it is clear that the government has taken on board the need to create a more unified, national market, in goods and services if not in labour. Changes in this direction would clearly give additional impetus to overall growth.

The future direction of regional development policy is now the subject of debate within China, a debate which takes in many of the key controversies facing planners. The draft Eighth Plan is ambivalent on the important question of whether a degree of formal recentralization of decision-making and resource allocation powers is now required, especially if the key reforms of the price and management structure of industry are to be pushed forward. It also leaves open the question of how the fiscal relationship between the Central Government and the provinces is to be reformed.

NOTES TO CHAPTER I

1/ Since the early 1980s, it has been official policy to seek to limit population growth in order to ease serious pressures on land and other resources. A campaign to limit the number of children to one per couple has had some success, mainly in urban areas, but tends to meet resistance in rural areas where the new policies have reinforced traditional values. It had been intended to hold the population to 1.113 billion in 1990 and to 1.2 billion by 2000. But the 1990 census revealed a total population of 1.134 billion and the annual average increase in the population, of 1.54 per cent between 1985 and 1990, was well in excess of the 1.27 per cent target. The revised goal is now to hold the average natural growth rate to below 1.25 per cent. This will be arduous in view of the fact that some 13 million women a year will reach child-bearing age in the early 1990s, apparently nearly twice the number of recent years.

2/ The role of township and village enterprises is discussed in detail in Chapter IV.

3/ This still predominantly relies on the traditional, socialist concepts of the gross output of industry and agriculture and, even as it is moving towards the System of National Accounts (SNA) method for recording economic activity, still underrecords the non-material product aspects of the economy. For technical notes on Chinese statistical terms, see Annex B.

4/ Savings rate data presented in this paragraph represent World Bank estimates.

5/ In 1985 (the latest year for which data have been released) the Central Government collected 38 per cent of total revenues and spent 45 per cent.

6/ Chinese statistics for inward investment from Taiwan Province are not systematically presented, but the amounts attracted to Fujian province are indicative and there is abundant anecdotal evidence to suggest that investment from Taiwan Province has continued to boom even when, in 1989, foreign investor interest in China waned somewhat due to a perceived increase in political risk.

7/ Control over wages has been difficult to implement and the tight credit policy was relaxed late in 1989, as discussed above, in the face of recessionary conditions in industry and rising unemployment.

II

POLICY FRAMEWORK FOR INDUSTRIAL DEVELOPMENT

A. THE APPROACH TO INDUSTRIAL STRATEGY: BETWEEN PLANNING AND MARKET ORIENTATION

Like other socialist countries that have experimented with market-oriented reforms, China has been faced with difficulties and risks when attempting to extend the role of the market to major industrial enterprises. Having been developed under the full protection of a command economy, many industrial enterprises are ill-equipped for the transition to hard budget constraints and to producing for the needs of the market. The inflationary burst in 1988 and 1989, and the measures taken to quell the rise in prices, have worsened the financial structure of many industries, by increasing their indebtedness.[1/] They face rising costs which can only partially be passed on to consumers of their products and the State has been obliged to bail them out, with worrying consequences for budgetary and monetary balance.

The long-term objective of industrial policy in China is to put productivity growth on a sustainable path, to replace outmoded technologies and lower the incremental capital-output ratio in general, concentrating the effort initially on the key producers of major capital and consumer goods. The large-scale investment in industry during the 1980s has had mixed results. Although the average age of one-third of all installed equipment had fallen to five years, as long ago as 1985, the general level of technology has remained low as factories have kept obsolete technologies. The complexity of investment decisions, often involving piecemeal additions of modern equipment to old plants generally lowers the returns from investment. The level of skilled technical personnel in the workforce remains far too low, as the losses inflicted on the skill base during the Cultural Revolution have yet to be made good. This calls for further changes in the industrial structure and the introduction of a more market-driven regime.

The stress is currently on a gradualist, non-radical approach to industrial reform, on consolidation of changes already introduced and on an industrial policy that avoids putting severe pressure on the price level. Price reform is to proceed cautiously, but price control, implicitly by administrative means, will not be abandoned while scarcities persist.

The ownership structure of that part of industry currently in State hands will be maintained and efforts to resolve the contradiction between State ownership and incentive-driven management will focus on attempts to separate ownership from operational management. No large-scale privatizations of the kind mooted by more radical reformers and placed at the centre of the economic agenda in eastern Europe are foreseen. Such a transformation of the socialist base of the economy seems unacceptable to the present leadership, even if they considered it practicable.

The agenda set by the government involves a *de facto* extension of its powers of resource allocation and an increased share in the revenues accruing to the State from taxation *vis-à-vis* provincial and lower level governments. A key aim is for the Central Government to use its enhanced powers to break down barriers to internal trade and pursue its commitment to the development of those areas of China which have fared less well than the now rapidly growing coastal provinces. It will concentrate its spending on infrastructure, energy, agriculture and selected areas of resource extraction and industry.

The output targets set by the Eighth Five-Year Plan (1991-1995) and for the ten years to the end of the century are generally modest by the standards of the reform decade. They are likely to be met and in many cases surpassed, provided that resources can be mobilized effectively. Whether the less tangible goal of pushing forward with piecemeal reforms to bring about a more efficient, integrated industrial structure without more radical liberalization than currently envisaged - of the ownership structure, of the labour market, of price-setting and decision-making, of the financial markets - can be achieved and how long this process will take are more questionable. The current consensus in the leadership suggests a highly risk-averse approach to economic management in general, which may sacrifice some growth. On the other hand the Chinese leadership can point to economic and social chaos as the risk attending over-rapid dismantling of existing economic structures.

The degree to which the Central Government will in fact be able to effect recentralizing policies and the extent to which it will be able to manage demand over a sustained period is difficult to gauge if it is unable to secure the full support of those localities that have benefited most from the decentralizing policies of the 1980s. Industrial policies of dispersion, reinforced in the 1960s and 1970s, have led to the spread of industry to some of the poorest provinces. The industrial heartlands of the east and north-east fell behind in terms of industrial output growth and share of investment until the new opportunities afforded by the reforms allowed the old industrial heartlands and the newly industrializing provinces along the coast to resume buoyant growth. In these provinces too, agricultural growth was more rapid during the 1980s and inequalities, measured in terms of income per head from agriculture, were allowed to widen, although urban incomes remained more uniform. If regional balance were sacrificed for the sake of growth, the backward regions in the centre, south-west and west would stagnate without being able to establish their comparative advantage in terms of cheaper labour and resources, while the coastal region would become increasingly dependent on external markets, not just for exports, but as the source of inputs to its industries.

China remains a country of relatively small industries, and decentralized industrialization policies have meant that industrial concentration ratios are low, other than in the steel and basic chemical industries. Despite the changes thrown up by reforms, capital has remained immobile as a result of the underdeveloped nature of financial markets and because of the priorities of local governments at the provincial and sub-provincial level.

The mushrooming of small-scale enterprises under various forms of ownership has accounted for much of the dynamism in the industrial sector in the late 1980s, when State-owned industries have become increasingly debt-ridden. There are many small enterprises with subcontracting ties with large State-owned enterprises; in the rapidly growing province of Guangdong, many small firms produce under contract for firms in Hong Kong. Other small-scale enterprises have sprung up to take advantage of opportunities to manufacture for local markets, producing building materials and other items difficult to transport. There are also many firms engaged in the manufacture of light consumer goods, whose marketing is carried out by sales representatives crossing the country.

In the 1980s, the proportion of output sold on the market and of inputs bought on the market has risen considerably, although the State still controls allocation of most key inputs and distributes the output of products in shortest supply, such as high quality rolled steel. The shortage of these key inputs, exacerbated by the very rapid growth rates of output during the 1980s, has created a considerable black market in goods whose legal prices do not clear markets.

Chinese industrial policy is thus currently entangled in rather problematic issue of the pace and direction of economic reform. Policy is at a crossroads in many respects; the choice lying between the more risky (because potentially inflationary and disruptive to employment) option of pressing ahead on the basis that the current state of State-owned enterprises is unsustainable, not least because of the burden of subsidies which is displacing other forms of expenditure, or adopting a cautious approach, in line with the preferences of more conservative elements in the leadership. The current thrust of Chinese macroeconomic policy, with its stress on the need to slow the pace of growth in output and control demand, is cautious and conservative in tone; although reform is still identified as the goal of policy, it appears to have taken on a more restricted meaning than in its heyday in the mid and late 1980s.

B. CONTOURS OF THE EIGHTH FIVE-YEAR PLAN (1991-1995) AND THE TEN-YEAR DEVELOPMENT STRATEGY (1991-2000): AN APPRAISAL

The contours of the Eighth Five-Year Plan and the Ten-Year Development Strategy envisage a much slower rate of overall growth than was achieved in the last Plan and particularly a slower rate of growth in gross industrial output at about 6.5 per cent annually. How this is to be apportioned between the various sectors is not specified, although it appears that rural enterprises are expected to continue growing more rapidly than State-owned enterprises (SOEs).

While output of priority products and intermediate goods in short supply will be encouraged (see Box II.1) to grow by the allocation of investment funds, special access to inputs and so forth, there are certain branches of industry whose output has grown too rapidly and has been of poor quality, exacerbating pressure on scarce resources. Hence the allusions in the various Plan proposals to the need to restructure the processing sectors of industry.

The township and village enterprises (TVEs), have grown very rapidly, providing jobs on a major scale (and accounting for about 10 per cent of net industrial output in 1988) but also competing with the overmanned and inefficient State sector for inputs and access to infrastructure. They therefore bore the brunt of the austerity programme of 1988-1989, when perhaps 1 million of them closed down. There is some evidence to suggest that a significant portion of the large increase in individual bank deposits seen in 1990 arose because collective and individually owned enterprises concealed their funds in private accounts. In fact, despite the currently conservative thrust of some aspects of policy, the leadership has recently been at pains to stress the important role played by the non-State-owned sector in industry.

Other themes in the Plan proposals stress the need to improve the efficiency of the State-owned sector and it is clear that the most pressing task is seen as being the raising of productivity by investment and by more efficient deployment of labour. While the need for such improvements is clearly understood, it is hard to see how radical improvements to the efficiency of China's State-owned industry can be effected without significant restructuring and reform of many aspects of the industrial environment. Nevertheless current policy, which, as discussed above, is cautious and gradualist in tone, does not seem to envisage drastic change.

It has often been argued, by economists outside China and by those inside the country who favour radical and speedy change to the economic system, that the growth of productivity in SOEs is hampered by the ambiguities surrounding the question of ownership.[2/] The new Plan sidesteps the issue by calling for further reform of management of industry and adopts a conservative stance on ownership. Limited experiments to date with forms of employee shareholding are apparently not to be extended to SOEs. The uncertainty over ownership relations is a powerful hindrance to the development of a more efficient industrial sector; it weakens incentives and complicates cross-boundary expansion.

A modest 6 per cent average annual increase in gross national product is envisaged throughout 1991-2000 (see Box II.2). On the other key question, that of price reform, a cautious and

Box II.1. Tenets of development goals for the 1990s

- Strengthening agriculture, basic industries and infrastructural facilities;
- reorganizing, transforming and upgrading the processing industry;
- according priority to the development of the electronics industry;
- actively developing the building industry and tertiary industry;
- rationalizing the economic structure and gradually modernizing it;
- building fewer new projects;
- reducing wasteful use of scarce resources;
- preserving public ownership as the main form and "appropriately" developing other forms; and
- selecting a number of large- and medium-sized backbone enterprises and key products as targets and breakthrough points for modernization.

Investment priorities: State-owned enterprises will invest 65 per cent of the total (up from 50 per cent in the Seventh Five-Year Plan) and their capital construction (building of new facilities) will grow less rapidly than technological transformation. In capital construction, priority will be given to agriculture, water conservation, energy, telecommunications and production of key products in short supply. In modernization, priority will be given to energy conservation, materials in short supply, improved quality and range, export products, import substitution, and industrial safety. Priority will also be given to some backbone enterprises and to the 'old' industrial cities of Shanghai, Tianjin, Shenyang, Wuhan, Chongqing and Harbin, relatively starved of investment for many years.

Box II.2. Targets for 1995 and 2000

The proposals put forward by the Central Committee of the Government for China's Eighth Five-Year Plan and Ten-Year Development Strategy are aimed at:

- maintaining an average annual growth of 6 per cent in gross national product throughout 1991-2000.
- boosting the share of service industries (i.e., wholesale, retail, banking, foreign trade, etc.) in GNP from about 25 per cent in 1990 to about 33 per cent in 2000;
- raising annual grain output from 435 million tons in 1990 to about 450 million tons in 1995 and to 500 million tons by 2000;
- increasing coal production from 1.08 billion tons in 1990 to 1.23 billion tons in 1995 and to around 1.4 billion tons in 2000;
- expanding crude oil output from the 1990 figure of 138 million tons to 145 million tons in 1995;
- raising total electricity generation to 810 billion kW in 2000, from 618 billion kW in 1990;
- boosting steel production from 65.8 million tons in 1990 to 72 million tons in 1995 and more than 80 million tons by the end of this decade.
- increasing the output of chemical fertilizer from around 90 million tons in 1990 to 100 million tons in 1995 and some 120 million tons in 2000;
- raising ethylene output from 1.5 million tons in 1990 to 2.3 million tons in 1995 and about three million tons in 2000;
- expanding cotton yarn output from 24.5 million bales in 1990 to around 27 million bales in 1995 and 31 million bales in 2000;
- boosting chemical fibres production from 1.6 million tons in 1990 to about two million tons in 1995 and 2.6 million in 2000; and
- increasing sugar output from 5.2 million tons in 1990 to around 6.7 million tons in 1995.

gradualist approach is being taken. In early 1991, the authorities apparently felt confident enough about the underlying rate of price inflation to engage in further moves to raise the prices paid by urban consumers for staple goods. Indeed, the pressure of price subsidies on the State budget has become so intense that even those who might not favour price reform seem to now accept the necessity to reduce the burden to the exchequer. Gradual moves in the direction of raising prices of inputs and staples towards market-clearing levels can be expected when the inflationary environment is perceived to be relatively benign.

Without being specific about the means for achieving an increased share of total revenues for the Centre, the Plan states that the method of revenue-sharing will be changed to give a greater portion to the Central Government. Other aspects of the Plan also explicitly herald a higher degree of central supervision over key areas, such as investment and the allocation of credit, to be overseen by the Central Bank.[3/]

On regional development policy, there is clear continuity with the previous Plan. The more highly developed areas are to concentrate on more knowledge and capital intensive industrial production, and on exports (see Box II.3). Industries that consume large amounts of raw materials and inputs and involve relatively crude processes, as well as energy and mineral extraction, are to be concentrated in the interior (known as hinterland) provinces.

In order to ensure that the trickle-down effects of rapid growth - industrial and agricultural - flow westwards into China's interior the Central Government would need to reverse some of the decentralization that has gathered pace during the 1980s. It could, paradoxically, create a market in China by administrative means. To this end, the Plan proposals envisage further moves towards commercialization of markets, by the creation for example of wholesale markets, and markets for capital, technology, information, housing and labour. Internal trade barriers are to be eliminated.

Related to the integration of the Chinese market is the creation of a mechanism for enterprises to take advantage of regional comparative advantage by creating subsidiaries in the interior or forming various kinds of association and processing agreements on the Japanese model. Although envisaged before, moves towards the creation of industrial conglomerates on these lines have so far been very hesitant, although provinces and localities have formed various administrative "horizontal linkages" in the supply of raw materials from the interior to the coastal area.

Box II.3. Eighth Plan regional profiles

Coastal Region:
- Transforming traditional industries;
- redeploying projects with high energy and materials consumption to interior;
- strengthening infrastructure;
- enhancing the foreign exchange earning capacity of agriculture;
- developing exports and tertiary industry;
- strengthening inter-provincial economic ties.

Hinterland:
- developing resource exploitation, agriculture, animal husbandry, communications and transport;
- promoting knowledge- and technology-intensive areas in large cities; and
- enhancing the former defence-related industrial enterprises in the economic development of the interior.

Industries in rural areas do not need the approval of municipal or provincial authorities in charge of a given output, but of county officials responsible for industrialization. Managing departments therefore have monopoly over production within urban areas under their jurisdiction, but cannot prevent entry by counties. It is thus clear that the creation of a more integrated national market will require a bureaucratic upheaval on a major scale. Meanwhile, the only competition faced by many enterprises is from TVEs in those provinces where rural industrialization has made strides. The Plan proposals are not specific about the way in which this maze of regulations is to be cut through, although it is apparent that the increased power of the Central Government *vis-à-vis* local units will be a key factor.

Other prerequisites to the creation of a national market are that the infrastructure - transport and communications - can support it and that the Central Government can exert sufficient leverage over investment decisions to ensure that local development policies mesh with national ones.

C. INDUSTRIAL PRIORITIES AND TARGETS: IN PURSUIT OF AN EFFICIENT INDUSTRIAL STRUCTURE

Within the modest target set for overall industrial growth of 6.5 per cent annually over the next ten years, some sectors stand out (see Box II.4). Target growth rates for physical output of important industrial raw materials and inputs are generally modest (see Table II.1). It is at first surprising, in view of the fact that output of coal, which accounts for over 70 per cent of energy production, is set to expand by 2.6 per cent a year while electricity output is targeted to grow at 5-6 per cent. This is partly because of expected efficiency gains and also because there are plans to bring on stream some major hydroelectricity projects, and the two nuclear power stations under construction, at Daya Bay in Guangdong and at Qinshan in Zhejiang province.

Table II.1. Targets under Eighth Five-Year Plan (1991-1995) and Ten-Year Development Strategy (1991-2000), by industry branch
(Values at 1990 prices, and real percentage change, unless otherwise indicated)

	1990	1995	Per cent change a/	2000	Per cent change b/
GNP (Rmb billion)	1,740	2,325	6.0	3,110	6.0
Gross agricultural output (Rmb billion)	766	878	3.5	1,043	3.5
Gross industrial output (Rmb billion)	2,385	3,270	6.5	4,544	6.8
Tertiary industry	473.3	..	9.0	..	9.0
Grain (million tons)	435	450	0.7	500	1.4
Cotton (million tons)	4.5	4.7	1.2	5.3	1.6
Coal (billion tons)	1.1	1.2	2.6	1.4	2.6
Crude oil (million tons)	138	145	1.0	..	..
Electricity (million kWh)	618	810	5.5	1,100	6.0
Steel (million tons)	66	72	1.8	80	1.8
Rolled steel (million tons)	51.5	59.5	3.0	..	..
Chemical fertilizer (million tons)	90 c/	100	2.1	120	2.9
Railway cargo (billion tons)	1.5	1.6	7.8	1.9	5.6
Yarn (million bales)	24.5 d/	27	1.9	31	2.4
Chemical fibre (million tons)	1.7	2.0	4.6	2.6	5.0

Source: *Renmin Ribao*, 16 April 1991.

a/ Annual average percentage change, 1990-1995.
b/ Annual average percentage change, 1991-2000.
c/ Effective components of chemical fertilizers in 1990 was 18.80 million tons.
d/ The output of yarn in 1990 was 4.63 million tons.

Fig.II.1. Gross industrial output targets for 1995 and 2000 (Rmb billion)

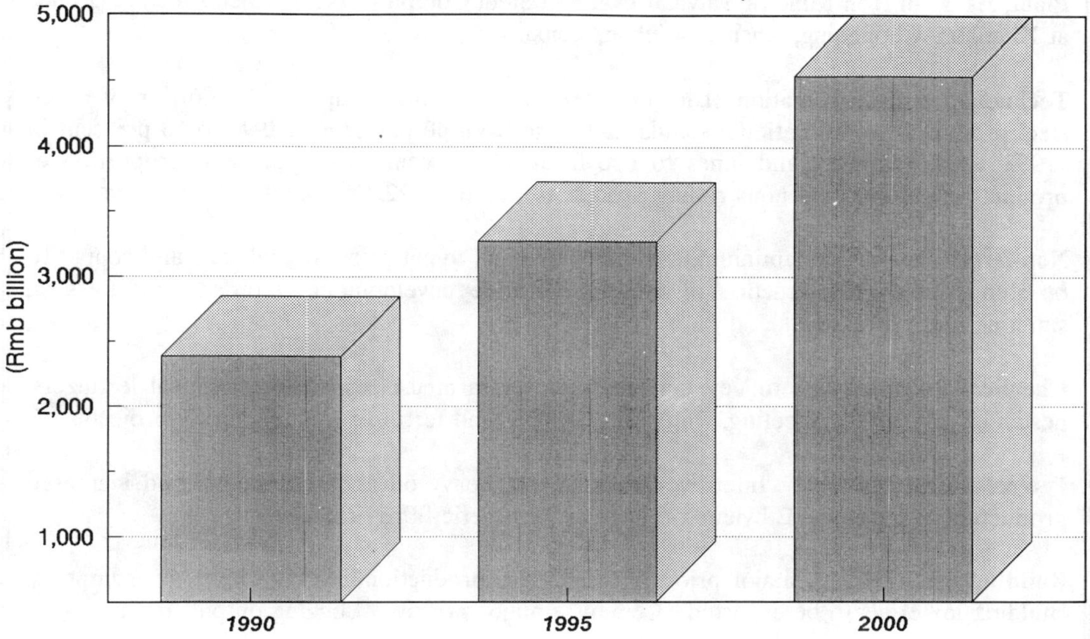

Fig.II.2. Selected physical output targets for 1995 (Million tons)

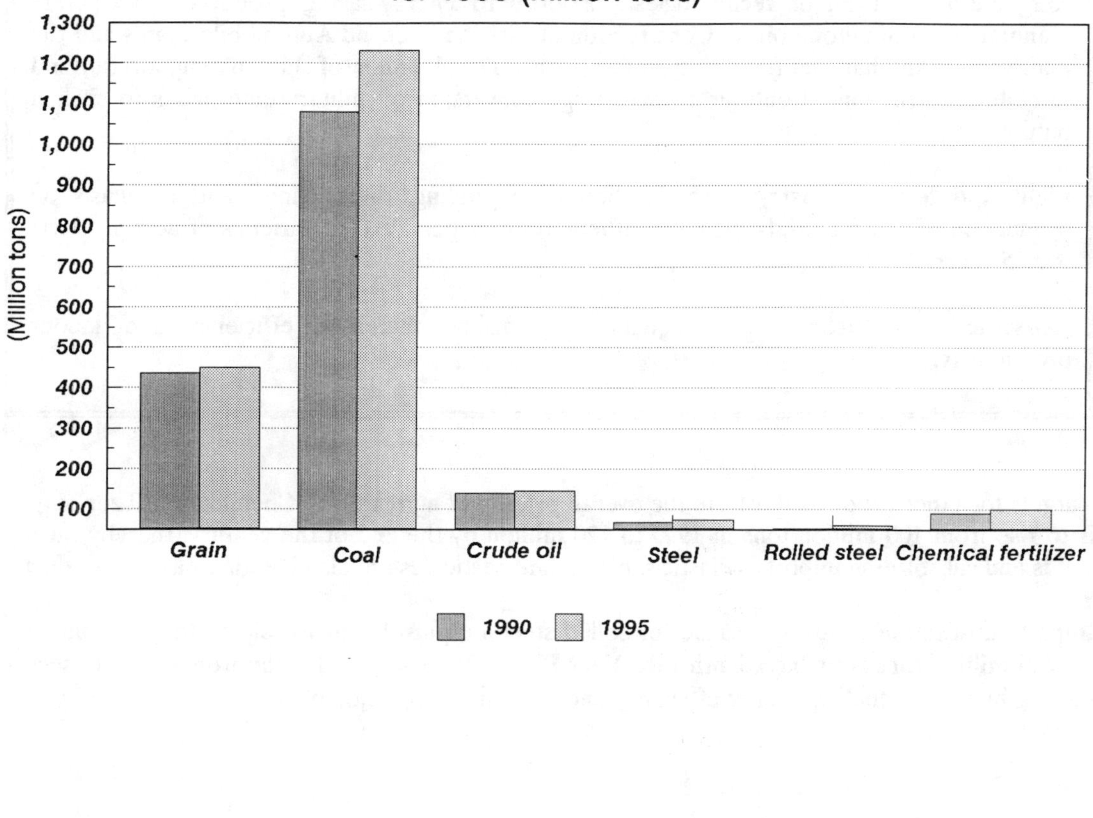

Box II.4. Eighth Plan industry profiles

Iron and steel: Continuing projects at Baoshan, Wuhan, Panzhihua, Maanshan iron and steel complexes. Other priority projects to be completed include Tianjin Seamless Steel Pipe Plant, Jianshan Iron Mine of Taiyuan Iron and Steel Complex. New projects including ones at Xuanhua, Chongqing, Anshan, Wuhan, Benxi.

Technological transformation of major enterprises to continue apace. Proportion of rolled steel produced to international standards to rise from 40 per cent in 1990 to 50 per cent in 1995. Ratio of plates and pipes to rise from 37 per cent to 40 per cent, proportion of products made by continuous rolling process to rise from 22.4 per cent to 35 per cent.

Non-ferrous metals: Aluminium, especially alumina, to get priority; lead, zinc and copper to be stepped up. Construction of mines to include development of on-site ore-dressing, smelting and processing.

Chemicals: Major task to develop materials for farm use, especially chemical fertilizers, pesticides and plastic sheeting. Integrated mining and fertilizer manufacturing projects.

Petrochemical industry: Intensive processing of heavy oil and residual oil and increased production of light oil. Ethylene projects and synthetic fibre projects.

Building materials: A major priority to increase production and cut energy consumption. Building materials to be exported. Cement a major priority. Logging output to rise.

Electronics: Identified as one of the key sectors. Integrated circuits and the computer industry of particular importance. Micro-electronics research and production joint groups, computer group companies, bases for the application, development and production of software; and bases for developing and producing telecommunication systems and equipment.

Machine-building industry: Systematically develop complete sets of advanced, efficient, and energy-saving machinery. Improve quality. Enhance capacity to develop and manufacture complete sets of major technologies. Priority to energy-saving products. Accelerate manufacturing of automobiles. Construction of First and Second Automobile Plants and the joint venture in Shanghai (involving Volkswagen). Development of shipbuilding, aircraft and satellite construction. Continue to encourage conversion of military enterprises to civilian output.

Light and textile industry: Acceleration of upgrading, diversifying and emphasis on production of raw materials such as synthetic fibre, paper. "8,000" varieties of new products over 5 years.

Construction industry: Strict quality supervision, increased efficiency and labour productivity.

Targets for other products illustrate the overall priority of agriculture. Chemical fertilizer output is to rise from 100 million tons in 1990 to 120 million by the end of the century, thereby raising yields and substituting imports. Plastic sheeting and pesticides are also identified as key products.

Import substitution is also envisaged for rolled steel, output of which is slated to rise from 51.2 to 59.5 million tons over the Eighth Five-Year Plan. Other targets for the iron and steel sector involve increased quality, energy efficiency and extensions to major plants.

Electronics has been identified as one of the priorities of the modernization effort and plays a major role in the Plan. Micro-computer output of 300,000 units is the target for 1995, by which time China is to produce video-recorders. The priorities for technology identify electronics as a vital area as China seeks to emulate the development path of the successful Newly Industrializing Economies (NIEs) by investing heavily in the application of computer technologies to industry as well as transport and communications.

In light industry, the emphasis is on continued import substitution (sugar, synthetic detergents) but above all on exports. With textiles still the single largest earner of foreign exchange the Plan calls for significant increases in the output of yarn and chemical fibres.

The target for automobile output, 900,000 vehicles by 1995, calls for a massive rise of 12 per cent a year and is mainly import substituting. Joint-venture production will play a key role in this increased output.

Outputs of raw materials for the construction industry, which have been in chronically short supply throughout the 1980s and have seen price rises well above the average, are targeted at fairly modest rates - 2.5 per cent a year for cement. This reflects the determination of the government to hold back the rate of growth of investment demand and to halt what it regards as wasteful construction of hotels, public buildings and other monuments by local governments. Moves towards commercialization of the urban housing market are envisaged, which would powerfully stimulate demand for construction materials, while at the same time serving to rationalise the structure of State-owned industry and raise the urban savings rate. The construction industry in general is expected to become much more efficient during the next five years, having performed poorly in recent years. In 1990, output value was down 3.9 per cent, area constructed fell 8 per cent per head, productivity slipped 1.5 per cent and losses increased. Careful monitoring of major projects by the central authorities is envisaged and an increased number of important projects are to be put out for competitive tender.

More significant than the physical output targets is the stress on all-round exploration (see Box II.5), upgrading and greater efficiency in energy and raw materials consumption, which would raise labour and capital productivity. Perhaps the most important theme lying behind the Plan proposals is the aim to create a balanced industrial structure. To achieve this, the question of resource mobilization is crucial.

D. RESOURCE ALLOCATION: CHANGING PRIORITIES

During the 1980s there were persistent attempts by the Central Government to constrain the growth of so-called extra-budgetary investment by SOEs and by the many forms of enterprise which sprang up outside the sphere of the Plan and the quota. This was necessary because the resultant excess demand for investment goods put severe pressure on prices and exacerbated the many bottle-necks which hold back the efficient growth of the Chinese economy.

It was not until 1989-1990, however, that a combination of the wholesale cancellation of many projects (deemed unnecessary) and macroeconomic policy moves (higher interest rates, higher prices for inputs, tight credit policies) squeezed domestic demand hard enough to bring about a fall in both the nominal and real growth of investment outlays. The government now possesses a combination of administrative and macroeconomic policy tools which allow it to modulate the growth of investment.

The current Plan calls for significant investment in resource exploitation in the poorer, hinterland provinces which are rich in resources, especially in minerals and energy, a policy which was in place during 1985-1990. It also specifies the annual growth in credit, holding it to 12 per cent a year in nominal terms. An increase in the Central Government's share of total revenues will permit some shift in resource allocation away from the more developed parts of coastal China, which are anyway better placed to attract foreign investment, towards less developed regions.

Box II.5. Fuel, power, transport and communications in the Eighth Plan

Coal: Expansion of large open cast mines in Inner Mongolia, completion of projects in Shanxi, Shaanxi, north-east, east and central south China. New projects in Shaanxi, Ningxia and Shanxi.

Electricity: Emphasis on hydroelectricity. Construction of thermoelectric plants near coal mines and transport nodes. Completion of second phase of Qinshan nuclear plant.

Oil and gas: Policy of stabilizing output from the east (Daqing, Shengli and Liaohe) and stepping up exploration of west, especially Tarim Basin.

Railways: Emphasis on coal-hauling main lines, a new north-south main line, and main lines in the north-west and south-west. Double-tracking and electrification. Encouragement to projects jointly financed by centre and local governments.

Roads: Major trunk roads, Beijing-Canton, Beijing-Shanghai, Shenyang-Harbin, Lianyugang-Lanzhou, express and motorways in busy coastal areas, help for poor areas in road construction.

Waterways: Coastal ports, especially hub ports along the main north-south sea routes. "All-out efforts" to develop inland river transport.

Airways: Emphasis on basic facilities, improving flight and air-traffic systems. Operating capacity to be increased.

Post and telecommunications: Accelerate automation of long-distance calls, Shanghai-Fujian, Fuzhou-Canton, Beijing-Shenyang-Harbin optical fibre trunk lines. Number of telephone exchanges to increase by 15 million lines, city telephones by 10 million lines.

The method of project approval and monitoring is also to be altered, presumably decreasing the authority of local level governments. Dedicated funds are to be earmarked for investment in energy and transport projects, both sectors which will receive priority access to investment funds. These funds are not to be diverted to other purposes and the authority of planning departments is to be enhanced so that, for example, ministries will not be permitted to engage in planning decisions without the participation of the Planning Commission.

Within the State Plan, investment by SOEs will take a greater share of total capital investment (65 per cent compared with 50 per cent in 1985-1990), which is slated to rise by a real 5-6 per cent a year over the period 1991-1995, far slower than previously. Of the total to be invested by SOEs, technical upgrading will advance more rapidly than capital construction and account for a larger proportion of the SOE investment total than under the Seventh Plan.

The Eighth Plan proposals state the need to reduce the number of new projects embarked upon, and it is hoped that a more realistic rate of investment will help solve the problems of cost overrun and slow completion which were endemic to investment projects in the 1980s.

A shift in regional focus is implicit in the Plan. While backward regions receive priority, several "old" large cities will be accorded importance in investment. Prominent among these is Shanghai, which has been discussed above. Also mentioned are industrial cities such as Wuhan and Chongqing whose location in middle China, along the Yangtze from Shanghai, suggests that the government is aiming to create dynamism in this large region, as it did further south-east during the 1980s.

These cities, along with Tianjin in Hebei, Shenyang in Liaoning and Harbin in Heilongjiang, have become increasingly congested and their infrastructure has deteriorated. Living standards in terms of housing and other facilities are falling behind, but they are rich in human resources and have relatively well-educated workforces. There seems to be a political trade-off discernible in the decision to allow them to invest more as well. Tianjin and Shanghai in particular have expressed frustration that the Central Government's overdependence on them for revenues has left them without the financial resources necessary to invest in local infrastructure and hindered their ability to attract foreign investment. Chongqing, in eastern Sichuan province, is the industrial heart of the province which has the largest population, but whose distance from the coastal region and modest share of investment outlays has left it among the slowest growing of China's provinces, despite being the home of many of the experiments that ushered in the reform era.

E. INVESTMENT MANAGEMENT: NEW EXPERIMENTS

Investment, despite its recent falls, still accounts for a very large share of total demand in China, and its growth is set to outpace that of consumption (3 per cent a year in national income terms) and even production (5 per cent) during the current Five-Year Plan period. As noted above, investment priorities have shifted to favour agriculture and "basic" industries under the Eighth Plan. Whether its targets are fulfilled will depend at least partly on the success of investment management.

The direct influence of the Central Government over economic decision-making and the proportion of outlays which it disposes of have diminished with reforms. By 1987, the number of materials subject to central allocation had fallen to 20 by 1987, when only 47 per cent of finished steel, 26 per cent of lumber, 16 per cent of cement and about half of coal was centrally allocated. Local governments have assumed allocative responsibility for some of the output no longer under central control, while enterprises themselves market a significant proportion of their output. Electricity is allocated by local governments.

Similarly, the government's share of enterprise revenue has fallen, with the Central Government's stake declining more steeply under decentralization. The result of the decentralization has been to make implementation of investment programmes more difficult; under both the Sixth and Seventh Plans the number of projects actually completed fell far short of target while the amount of investment exceeded planned levels by significant sums. Funding through the State budget accounted for only 8.9 per cent of investment outlays in 1989, down from 16 per cent in 1985, with bank loans and retained funds expanding significantly to fill the gap.

The Central Government has, in response, cut down on the number of projects it undertakes and drawn in matching funds from local governments to fund projects. To those projects which it has identified as key projects, it has accorded generalized priority and it has carefully monitored their progress. This policy can be expected to continue with the Central Government's spending concentrated on agriculture, energy, transport and so forth in combination with efforts directed at a recentralization of finances.

Local governments, meanwhile, seek to increase their revenue base and invest in projects which will foster growth and create employment. The main source of local government revenue is the profits of its own enterprises, which are sometimes supplemented with "extra-budgetary" levies for specific purposes. Local governments have a powerful incentive to invest when they can accrue revenues by developing industries using locally produced raw materials, whose underpricing makes the establishment of processing industry an attractive option, even when the industry is inefficient.

Measures taken in 1989-1990 have done something to realign the cost/benefit structure of investment decisions, at least as far as enterprises are concerned, the decline in their own profitability being an important factor. Credit control and interest rate policy will remain the chief means of regulating investment outside the direct sphere of the Central Government and the expected recentralization of finance should lead to more careful control of new projects.

In furtherance of a more rational investment programme, the Central Government will also, step by step, seek to readjust the revenue-sharing contracts which it signs with enterprises (and local governments), both to increase revenues and to influence the investment climate. Loan payments (principal and interest) are currently fully deductible and there will be experiments to change this as part of an effort to subject enterprises to a less "soft" budgetary constraint.

F. FOREIGN DIRECT INVESTMENT: RISKS AND OPPORTUNITIES

The "Open Door" policy, in place throughout the 1980s, has led to the absorption of significant, although by no means enormous, amounts of foreign direct investment and has fostered the creation of many thousands of jobs in processing industries, often small-scale, low technology operations serving light industry in Hong Kong. The prosperity of Guangdong province in particular, and of other areas in the coastal region (such as Fujian province, which has close links with Taiwan Province) has increased markedly as a result. Export revenues, which reached an estimated $60 billion in 1990, have soared, from $25 billion in 1985. Of course this export growth has had an import content, as evidenced by the very large flow of re-exports which enter and leave China via Hong Kong, but the average trade deficit in the five years 1986-1990 has been easily manageable, at $3.1 billion a year.

As well as export promotion, a major motive for seeking to attract foreign direct investment has been the desire for access to high technology manufacturing processes. A number of large-scale projects have been agreed, involving major Western and Japanese companies in joint-venture production of products ranging from automobiles to coal-mining and mineral extraction, using advanced technology and substituting imports.

The first sector of the Chinese economy to be opened to foreign investment on a large scale was offshore oil exploration. A total of 27 contracts was signed following the first round of bidding in 1982 and 24 companies bid in the second, more favourable, round in 1985, despite the disappointing results from the first few years of offshore exploration. Foreign companies are now being encouraged to participate in onshore exploration.

Foreign participation in the Chinese economy takes various forms, ranging from wholly-owned foreign ventures to joint ventures, contracting and processing agreements. Although the precise terms available vary according to location and the nature of the project, the packages of exemptions from taxation, repatriation of profits, exemptions from import restrictions and priority access to labour and inputs are generally competitive with terms offered by other developing countries, and the taxation regime was updated as recently as April 1991 at the National People's Congress session which approved the current Five- and Ten-Year Plans.

Until 1988, investment commitments were running in the range of $4.5 billion a year on average. Since then the pace of inward investment has slowed in the face of ever more intense competition from countries in the region whose resource endowment makes them more attractive, the waning of the initial euphoria about China from foreign companies, especially those from the United States and Japanese, and a heightened perception of political risk after 1989. In future the pace of inward investment will depend on improving the investment environment, especially the infrastructure and bureaucracy, and on the world economic climate. With the opening of eastern Europe and the USSR, as well as an improved economic climate in several countries in Latin America, China faces increased competition for all forms of investment inflow in what may be a period of capital shortage as the Japanese and especially German current account deficits shrink and the financial problems of banks and large companies in the United States and other countries impose constraints.

From the perspective of foreign companies China has the potential attraction of an enormous domestic market. However, shortages of inputs are acute, despite preferential access, and much of the input to manufacturing must be imported for quality reasons, posing logistical problems. Construction of plant has been inordinately protracted, partly due to a shortage of skilled site surveyors and other contractors to the construction industry, and projects have run into significant time and cost overruns before coming on stream. The services sectors of the Chinese economy, especially banking and finance and tourism ventures, are potentially attractive to foreign investors, involving lower entry costs, but access to them has not been significantly liberalized. The Chinese have acknowledged the majority of these difficulties and steps have been taken to improve the investment climate. The successful completion of the Pudong zone of Shanghai could significantly add to the attractiveness of China as an investment location for large-scale projects.

The current policy stresses the need to continue to seek foreign investment and recognises the benefits of the small-scale investment from Hong Kong and other Chinese communities abroad. At the same time the Plan proposals call for the establishment of more "open" cities on landlocked borders aimed, presumably, at increasing economic ties with the USSR and the Republic of Korea, as well as with neighbouring countries. It also endorses the performance of the Special Economic Zones and open cities already operating preferential policies, and policy is thus in continuance with the previous two Plans, with the chief aim being to attract high-technology, export-oriented investment.

G. THE TECHNOLOGY DILEMMA: ISSUES AND OPTIONS

The goals of the modernization drive commit China to seek to create an industrial base producing to the most advanced technological standards. It has already made impressive strides in many areas of technology, as symbolized recently by the successful launch of communications satellites, as well as the development of sophisticated defence technologies.

Technology is a major theme underlying the Plan proposals, which set forth eight areas for scientific and technological research over the period to 1995. Goals are also set for the proportion of output of certain sectors - machine building among the important ones - which should reach advanced levels by 1995.

While the upgrading of plant and equipment and the building of advanced industrial complexes is an important way of increasing the productivity of the economy, there are associated problems. The first concerns the incremental efficiency gains from piecemeal additions of modern equipment to existing plant. As discussed above, the nature of enterprise investment decision-making encourages this approach. Many plants combine equipment of widely varying age and efficiency in one rather chaotic production line while continuous mass production techniques have yet to be implemented on a large scale. There might be greater returns from the purchase and installation of complete, possibly second-hand, production assembly lines. A more rational and systematic approach to technological upgrading depends, like so much else, on reform of the whole enterprise management and financial system as well as guidance from a more assertive Central Government in the absence of market signals.

There is another long-range consideration relevant to the appropriate level of technology for a country with so large a surplus of labour. The Eighth Plan calls for the creation of 32 million urban jobs (6.4 million a year) to hold the urban unemployment rate below 3.5 per cent. The labour force is growing by about 11 million a year and there is already a surplus labour force estimated at 200 million, swelled in 1989-1990 by the closure of many TVEs. The skill level of this labour force is also a concern.

If parts of industry are to adopt advanced technology, and if enterprise reform proceeds to the point that enterprises are not obliged to hoard surplus labour, much of the unemployment currently disguised could be forced into the open, with worrying political and social consequences. These considerations are an influence on the question of technology and on the appropriate growth rates for production as opposed to the services sectors. The government is clearly hoping that the rapid development of labour-intensive tertiary industry and of rural manufacturing will absorb China's surplus labour, allowing key State-owned enterprises to adopt advanced technologies without effectively depressing the productivity gains. This will take many years and meanwhile the key to technological progress that optimises productive efficiency will retain its selectivity (see Box II.6), while the economy would clearly benefit from large-scale technological upgrading of the physical and communications infrastructure.

Box II.6. The role of technology in the Eighth Plan

Agriculture:	Enhancing the role of agronomy in terms of planting technology, disease and pest control, animal husbandry and aquaculture, storage and processing, forestry.
Energy:	Increasing yields from eastern oilfields; and enhancing energy efficiency.
Communications:	Priorities include railway efficiency, better materials for road construction, navigation for civil aviation, design and manufacture of aircraft.
Raw materials:	Technologies to produce catalysts for chemical industry, liquefaction, various energy-saving applications.
Microelectronics:	Priority areas are 3-micron, 1-micron and sub-micron, special integrated circuit and other special equipment, optical communications, remote sensing, large computer systems and software.
Other:	Other technologies to focus on population control, disease, pollution control, monitoring and dealing with natural disasters.

H. THE ENTERPRISE SYSTEM: REFORMING THE OPERATING ENVIRONMENT

The Chinese enterprise system is a mutated form of the Soviet model under which a centralized administrative system, divided into departments or industries organized functionally, embraced all production. At the apex of the whole system was the planning commission, the role of the enterprise being to carry out its instructions. While the Soviet system became highly concentrated, Chinese industrial development was much more decentralized, with local governments at various levels managing enterprises and production dispersed over many factories. Chinese enterprises also took on responsibility for a permanent labour force, with the tight restrictions on labour mobility and the dismantling of social security systems managed by labour unions in the 1960s. Despite the increased autonomy over some aspects of decision-making afforded by the reforms, Chinese State-owned industry remains largely under the control of Central Ministries and/or provincial industrial bureaux.

Industrial decision-making is therefore a highly complex process of negotiation. While the government relies on taxes from industry for its revenue flows, it has traditionally manipulated relative prices to ensure that most of industry is profitable in accounting terms. This profitability has been greatly eroded, however, by the cost pressures mentioned above and declining profitability was exacerbated by the very depressed state of domestic demand during 1989-1990 which left the output of many enterprises unsold and produced a chain reaction of cash flow problems as banks could not recover their loans.

The level of profit retention by enterprises, an important source of investment and means of rewarding the labour force through various benefits, is the result of a bargaining process whereby revenues are divided under the contract management responsibility system (CMRS) (recently reinforced by the double guarantee system). This system involves the signature of a multiyear contract setting the amount of profit or tax to be delivered. The tax rate on revenues in excess of the contracted amount can vary from 100 per cent to zero.

The predominant form of remuneration for workers remains fixed wages plus annual bonus. Rules have been set determining the amount by which total wages can rise as a proportion of profits. Despite the nominal weakness of formal organizations representing the interests of workers, it appears that their *de facto* influence over the distribution of retained profits is strong and they have tended to favour egalitarian distribution systems, despite official endorsement of the principal of differential incentives according to performance (each according to his work).

The system of labour allocation, under which there is virtually no national labour market allowing the transfer of skills and, despite the formal enactment of laws replacing permanent employment with contracts, virtual life tenure is still assured, militates against incentive-driven rises in productivity. Chinese workers are also divided into very few wage grades and the *de jure* authority which managers have to promote 3 per cent of the workforce if targets are met is often not available *de facto*.

One of the factors inhibiting productive investment is the nature of the depreciation allowances which form the bulk of the enterprises' retained funds. These are calculated at relatively low rates, and the difficulty of disposing of outworn assets, as well as the flow of funds arising from retention of every piece of equipment, inhibits its replacement. Depreciation is also accrued in small amounts, encouraging the purchase of a single piece of additional equipment rather than integrated renovation of the production line. The underdeveloped state of the market also exacerbates shortages of inputs and encourages enterprises to devote a great deal of effort to tangential activities, such as in-house generation of electricity and repair and maintenance of plant and buildings, including manufacture of spare parts for key machines.

Other administrative impediments to efficient growth include the need for each large or medium-sized enterprise to have an organizational structure that replicates the nature of the administrative hierarchy to which it is subordinate, and the difficulty of diversifying, either geographically or into new products, because of the territorial and sectoral monopoly system.

The role of the enterprise as provider of housing, health and other services also helps to exacerbate overmanning, by loading the payroll with large numbers of employees not related to the production process. This lowers measured *per capita* productivity as well as complicating the task of management.

In the context of the highly restricted operating environment faced by industry, the State has encouraged the emergence of "enterprise groups". These are organized as affiliations of enterprises with production linkages, still under local or administrative control, and have not evolved into ways of diversifying or creating subsidiaries.

The above brief description of some of the factors hampering the development of an integrated, efficient industrial structure in China, should not obscure the fact that significant achievements have been made, notably during the 1980s, in developing and upgrading industry, including the selective adoption of some high-technology processes. The envisaged enterprise reforms during the Eighth Plan are presented in Box II.7.

I. MAJOR INDUSTRIAL PROJECTS: STRENGTHENING THE INFRASTRUCTURAL AND INDUSTRIAL BASE

The single largest project envisaged in the Eighth Plan is the development of the Pudong area of Shanghai. Costed at at least $20 billion, it would absorb nearly 4 per cent of the total outlay of Rmb 2,600 billion planned for investment in the five years to 1995, if it could be completed within five years. The amount the State will contribute to this major venture has not been specified and foreign funding is being sought. But it is likely to require substantial domestic financing.

Box II.7. Eighth Plan reform of the enterprise system

Large- and medium-sized State-owned enterprises to be the focus of reform. Enterprise Law to be implemented. Contract responsibility system to be renegotiated with base numbers more favourable to the State. A more standardized system to be implemented. Enterprises to be prevented from turning over too many profits to individuals. Trial implementation of separating profits from taxes, repaying loans after paying taxes (loan payments are currently tax-deductible), and contracting enterprises after paying taxes.

Depreciation rates to be increased to encourage efficient investment. Accelerate reorganizations, associations and mergers. Organize a number of competitive interregional and interdepartmental groups. Reform management, internal labour and personnel systems, wage systems, financial and other management procedures.

Markets:	Developing wholesale markets, capital, technology, information, housing and labour markets, and eliminating internal trade barriers.
Prices:	Maintaining control over overall price level; correcting the excessively low rate of State-fixed prices; raising grain procurement prices, reduce price subsidies. Gradually eliminating control over the prices of ordinary processed goods which are not in short supply; and eliminating the dual track pricing system for the prices of some means of production.
Financial system:	Raising the share of State revenues in national income; increasing the Central Government share of total revenues; maintaining current budget balance; encouraging development of the banking sector; experimenting with securities exchanges in large cities.
Wages:	Establishing a mechanism for normal increases in wages; linking total wages more closely with enterprise performance, and skill; reducing distribution in kind; and enforcing personal income tax.
Housing and social security:	Reforming the housing system by commercializing housing units; and establishing social security systems.
Planning and investment:	Adjusting the relationship between mandatory and guidance Plans to expand the role of market mechanisms in the economy; reforming the method of deciding on delegation of authority to localities to approve projects; centralizing management of key projects; dedicating special purpose funds to energy and transport; implementing measures such as the issuance of bonds to be adopted to raise construction funds; and allowing free play of tenders in project construction.
Labour:	Providing 32 million jobs during 1991-1995 in cities and towns amid efforts to keep urban unemployment below 3.5 per cent; developing labour-intensive industry, especially the services sector, and controlling the movement of labour to the cities; and encouraging rural surplus labour to stay in the locality, by vigorous development of non-agricultural industries in rural areas.

It is worth noting that one of the most controversial of all the projects mooted since 1949 is the plan to dam the Yangtze River at the Three Gorges. The project first surfaced in 1958 and would be on a massive scale, involving the world's biggest dam and hydroelectric power station, a 300 mile long deepwater channel along the Yangtze and the displacement of as many as 1 million local inhabitants. Its supporters point to the fact that it would add 17,000 MW (over 10 per cent of installed capacity), alleviating the acute electricity shortage in the Yangtze region, and would end floods. It had been announced in 1990 that the project would be shelved until 1995 at least, but it was reconsidered while the Plan proposals were being finalized. So controversial (and expensive) is the scheme, whose opponents argue against it on environmental and cultural grounds, as well as asserting that it would create silting of the upper reaches of the Yangtze and that it represents a dangerous concentration of resources - that the Plan proposals simply place the scheme under consideration. Meanwhile, as noted above, a slower industrial growth rate, greater efficiency in the use and production of electricity (the energy intensity of GNP is to fall by 2.2 per cent a year) and the addition of capacity on a smaller, piecemeal, scale are the basis of current thinking. Other major projects in the Eighth Plan are in iron and steel, non-ferrous metals, energy and mining, transport and communications and agriculture.

The decision to construct a large-scale iron and steel complex at Baoshan near Shanghai, the first integrated steel complex in the country, involving Japanese technology, was taken as long ago as 1978, a period characterized by very ambitious projects that later had to be scaled down. Baoshan was frozen in 1981 as part of the first of the readjustment phases of the 1980s, then resurrected, with the cancelled second stage revived in 1983. Investment in the first and second phases of the Baoshan plant was $6.4 billion, far higher than the original estimate of $2.2 billion for the whole project. The first stage was completed in 1985 (postponed from 1982) and the second stage is due for completion during 1991, bringing output from 3.7 million tons to 6.7 million (10 per cent of the total). The second phase of this giant project was undertaken despite the fact that the ore mixed in the plant has to be imported. The Eighth Five-Year Plan specifies continuing projects at Baoshan and at Wuhan (built with German and Japanese assistance), as well as at other steel centres: Panzhihua, Maanshan, Tianjin, Taiyuan. There will be new mines belonging to steel plants at Xuanhua, Chongqing and Anshan, and new plants at Wuhan, Meishan and Benxi.

Other key mining projects in the non-ferrous metals sector are planned, to increase output of alumina, lead and zinc and copper. Mining and fertilizer manufacturing projects in Guizhou and Hubei will raise chemical fertilizer output and there are projects identified in Yunnan, Jilin, Inner Mongolia, Shaanxi and Jiangxi.

In energy, open cast coal mines in Inner Mongolia are slated for expansion, projects in Shanxi, Shaanxi, Ningxia, and elsewhere are to be completed and new mines opened. Construction of hydroelectric plants will be undertaken in the upper reaches of the Yellow River, the Yangtze and the Hongshui and in various other places. Thermal power plants will be constructed and the second phase of the Qinshan nuclear plant in Zhejiang province will be undertaken. Crude oil projects will concentrate on exploration in the west and on attempts to stabilise the output of the major eastern fields - Daqing, Shengli and Liaohe.

Railway construction remains a major priority, especially now that the energy resources of far-flung provinces are to be tapped. Coal-hauling lines, a new north-south main line, and main lines in the north-west and south-west are identified, along with electrification and double-tracking projects. Trunk line roads are to be constructed, sea and river ports upgraded. In telecommunications, optical-fibre trunk lines are to be constructed, linking Shanghai-Fujian, Fuzhou-Canton and Beijing-Shenyang-Harbin.

Investment in agriculture will focus on afforestation, land reclamation and irrigation. The aim is to add 30 million mu to the area of farmland irrigated and to divert water from south to north. Two key water control projects are identified as the Xiaolangdi and the Wangjiazhai as well as a project to divert Yellow River water to Shanxi. As discussed above, the enormous Three Gorges project will remain under study. Details pertaining to twenty key projects are presented in Annex C.

NOTES TO CHAPTER II

1/ Industrial firms are becoming increasingly indebted. Changes in relative prices have eroded the profitability of many large firms producing key items at State-controlled prices. A portion of accounting profits have been redefined as taxes replacing the handover of profits to the State; higher interest rates and depreciation allowances have also trimmed profit, while costs are undergoing an upwards pressure from higher prices for raw materials, and wages and bonuses. According to a World Bank estimate, current liabilities (overwhelmingly to banks) as a proportion of total net worth in State-owned industry had risen from 11 per cent in 1978 to about 76 per cent in 1989. By 1990, when the State had to pump large volumes of credit into the State-owned sector, this proportion had risen further.

2/ The net worth of an enterprise has been divided into a "State" and "enterprise" share, with ownership of the assets attributed to the side which was the source of the investment; investment from enterprises' retained funds having increased significantly, and the enterprise "share" of its own net worth has also risen, contributing to an increasing uncertainty as to managerial authority, claims on income and assets involving the State, the enterprise, the banking sector and of course local governments.

3/ The Central Government has recently altered the basis for sharing foreign exchange revenues and will now have access to a larger proportion of the foreign exchange earned by exporting areas.

III

STRUCTURE AND PERFORMANCE OF THE MANUFACTURING SECTOR

A. GROWTH AND STRUCTURAL CHANGE

The need to correct imbalances

During the 1960s and 1970s, industrial output grew at an impressive 10 per cent a year. In the 1980s, growth accelerated to 13.5 per cent per annum. By the end of the 1980s, several decades of intensive industrialization had made China a leading producer of many industrial products. Its share of world industrial output more than doubled during 1977-1986 and its manufacturing value added ranked seventh in the world in 1987.[1/]

By the mid-1980s, China had become the world's largest producer of cotton yarn and textiles, as well as cement and sulphuric acid. Its output of steel ranked fourth in the world and a number of new and relatively technology-intensive product technologies had been mastered, often with external assistance. After several years of very rapid growth in the economy, driven by accelerating growth in industrial output and investment, the authorities brought about a drastic deceleration in the growth of gross industrial output in 1989, when it slowed from 21.0 to 8.5 per cent, further moderating to 7.6 per cent in 1990. This slowdown was caused by a sharp squeeze on credit, which cut the growth of demand in general, and by prioritizing the key, large-scale, State-owned industrial sector in terms of access to working capital and scarce inputs. As a result, the industrial sector, especially smaller-scale township and village enterprises, bore the brunt of the squeeze, while energy and raw material production were encouraged to grow.

The austerity programme, which was originally expected to last for three years when it was introduced in 1989, but had been largely abandoned by mid-1990, led to the closure of factories and lay-offs of over 15 million workers in State enterprise in the first quarter of 1990. The crisis which still faces industry is, in part, the legacy of central planning. State-owned enterprises and bureaux gained easy profits by reselling inputs allocated to the planned sector and rising prices for raw materials failed to give rise to a supply response from the State sector.

Long-term structural change

The structural evolution of the Chinese industry occurred in three phases. In 1957-1978, heavy industry grew more rapidly than light industry.[2/] In 1978-1983, this trend was temporarily reversed and light industrial output was emphasized as the State sought to make up for years of neglect of consumer demand. In 1983-1988, however, the traditional bias towards heavy industry reasserted itself, when energy and transport projects assumed greater prominence. The share of heavy industry in net industrial output rose from 54.1 per cent in 1981 to 61.2 per cent in 1985. Its share of net industrial output stood at 56.6 per cent in 1990, compared to around 58.0 per cent in 1988 and 1989. The growth rate of light industry is influenced by the growth of consumer demand. Private consumption grew at an annual average rate of 7.4 per cent in 1980-1988, after having been 5.2 per cent in 1965-1980, but the very much more rapid growth of investment (see Annex Table A-5) discussed above held the share of private consumption in GDP to 56 per cent in 1988, after having been 61 per cent in 1965. In general, there is evidence to suggest that demand for domestically produced durables has tended to stagnate, while the savings rate, especially in the rural sector, has been increasing. After an inflation-induced surge in demand for consumer goods in mid-1988, consumer demand has slowed down and was very sluggish in 1990, when the propensity to consume fell, partly due to a problematic political climate. Long-term growth patterns reveal growth rates across several industrial subsectors which have been spectacular by any standards (see Table III.1).

During the first phase of the reform period (1979-1982), light industry was the driving force behind rapid industrial output growth, largely due to the accelerated growth of small-scale enterprises lying outside the scope of State and plan. This expansion absorbed much hitherto underutilized productive capacity, in particular in the rural sector.

Since 1983, however, heavy industry has once again taken the lead. Rapid expansion of the capital goods branch of industry has been the result of soaring investment demand. The period 1983-1988 was of exceptionally brisk expansion, surpassed only by an annual average 18 per cent (from a much lower base). In 1989, the policy-induced contraction in investment (down 11 per cent in nominal terms [investment data for 1989 and 1990 are highlighted in Annex Table A-6]) led to sharp falls in the production of some consumer and capital goods. State-owned enterprises reduced investment by 9.2 per cent; investment by collective enterprises fell by 28.1 per cent; investment by private enterprises fell by 4.3 per cent. Among the capital goods sectors which were hard hit were motor vehicles, tractors and mining equipment (see Table III.I). Among consumer durables output, luxury items such as cameras, washing machines and refrigerators, fell in 1989. In 1990, net industrial output grew by 1.6 per cent, with the manufactures of refrigerators, machine tools, sewing machines, washing machines, timber, motor vehicles, cameras and bicycles recording very sharp declines.

The extent of growth and structural change within Chinese industry can be gauged from the physical output of major industrial products presented in Annex Table A-7. Over the past four decades, one outstanding characteristic of all five manufacturing groups has been the substantial expansion in physical output. There were some fast rising industries, such as chemical fibres, yarn, sewing machines, bicycles, wrist watches, television sets, cameras, etc. Most of the heavy industry products seem to have registered a steady expansion until 1980. The preponderance of heavy industry over light industry in the Chinese industrialization process is corroborated by physical output data presented in Annex Table A-7. What is most striking is that declining physical output values of heavy industries seem to have occurred with increasing physical output values of light industries, with a few exceptions during the early 1980s when manufacturing activities were constrained by severe energy shortages. This reflects not only the dichotomy between light and heavy industries, but also an uncoordinated import of technologies and intermediate goods for consumer durable industries.

Textiles and food manufacturing continue to dominate the structure of net industrial output value (see Table III.2). Among the heavy industries, the contribution of the machine building industry to net industrial output value was strikingly high at around 10.0 per cent in 1990, followed by chemicals and allied products.

The resultant problems of "run away" industrialization in most of the 1980s manifested themselves in an unbalanced industrial structure which did not allow China's resource endowment and productive capacity to sustain a robust pace of industrial expansion. The imbalances between the basic industry and infrastructure on the one hand and the processing industry on the other surfaced, and the slow development of the energy industry and infrastructural facilities could no longer support the rapidly expanding processing industry which suffered from unplanned and overlapping activities. As the pace of industrial expansion cooled down in 1989 in response to the rectification programme, policy-makers became optimistic about the future course of a prudent industrial modernization. The rectification programme aims at increasing the production of major farm products, easing the shortage of energy and raw material supplies and improving transportation facilities.

Table III.1. Growth of net output by industrial branches, 1960-1990, selected years
(Percentage)

	1960	1970	1980	1985	1986	1987	1988	1989	1990
Light industry	-9.8	18.1	19.0	22.7	13.1	18.6	22.1	8.2	3.9
Heavy industry	25.9	42.3	1.9	20.2	10.2	16.7	19.4	8.9	6.9
Total industry	11.2	30.7	9.3	21.4	11.7	17.7	20.8	8.5	1.6
Selected heavy industry products									
Pig iron	24.0	33.3	3.5	9.6	15.5	8.7	3.7	2.0	7.2
Steel	34.5	33.5	7.7	7.6	11.6	7.8	5.6	3.6	7.7
Cement	27.5	40.8	8.1	18.6	13.8	9.8	12.8	0.1	-0.3
Plate glass	11.7	12.6	18.4	17.9	5.3	11.6	25.7	15.8	-4.4
Timber	-8.6	15.2	-1.5	-1.0	2.8	-1.4	-3.0	-6.7	-11.9
Chemical fertilizer	52.3	39.2	15.6	-9.5	5.6	19.7	4.1	3.6	4.3
Machine tools	32.9	62.3	-4.3	25.2	-2.1	5.2	4.5	3.1	-24.7
Motor vehicles	15.3	64.2	19.7	38.2	-15.6	17.8	36.6	-9.5	-11.9
Tractors	156.0	154.0	-28.8	19.1	-7.6	42.7	27.2	-15.7	-1.0
Power generating equipment	39.8	43.7	-32.6	20.5	28.2	30.4	17.9	5.8	4.4
Mining equipment	9.6	56.3	-38.4	21.8	-4.4	-1.1	29.1	-15.1	-3.6
Selected light industry products									
Cotton cloth	-28.0	11.4	12.1	7.1	12.3	5.0	8.6	0.7	-0.2
Chemical fibres	96.3	51.5	38.0	29.0	7.3	15.5	10.7	13.8	11.7
Television sets			87.6	66.1	-12.5	32.5	29.5	10.4	-3.0
Refrigerators				164.5	55.7	77.9	88.8	-12.6	-31.0
Washing machines			1,261.1	53.5	0.7	10.8	5.7	-21.1	-19.7
Electric fan			210.7	17.8	11.1	3.7	22.8	11.0	16.2
Radio sets			117.6	-17.9	-0.7	11.0	-12.2	18.5	11.5
Watches	501.9	33.3	29.5	43.1	34.6	-16.0	8.5	9.2	14.8
Cameras			56.6	42.6	13.2	26.7	21.6	-21.5	-13.0
Bicycles	32.5	26.3	29.0	12.8	10.6	15.4	5.7	-11.2	-14.6
Sewing machines	35.2	22.4	30.8	6.0	-0.2	-2.0	1.4	-2.7	-20.4
Machine-made paper and paperboard	5.9	11.1	8.5	20.5	9.7	14.2	11.3	5.0	2.9

Source: State Statistical Bureau of China.

Table III.2. Distribution of net industrial output value, 1982-1990, selected years
(Percentage)

	1982	1985	1988	1989	1990
Net industrial output value	100	100	100	100	100
Light industry	43.68	33.8	42.37	42.2	43.4
Heavy industry	56.32	61.2	57.63	57.8	56.6
Total (all listed)	76.02	87.9	87.75	89.7	89.17
By branch of industry					
1. Food manufacture	)	3.5	3.88	3.9	3.94
2. Beverage manufacture	)	1.8	2.30	2.1	2.27
3. Tobacco manufacture	) 11.12	4.2	5.17	5.3	5.84
4. Fodder manufacture	)	0.1	0.26	0.3	0.28
5. Textile manufacture	]	9.1	9.47	9.7	9.67
6. Chemical fibres	] 12.44	1.0	1.19	1.2	1.51
7. Clothing	1.73	1.8	1.71	1.8	1.98
8. Leather, furs and manufactured goods	0.77	0.8	0.79	0.8	0.89
9. Timber processing, bamboo, cane, palm fibre and straw products	)) 1.12	0.6	0.61	0.6	0.47
10. Furniture manufacture	)	0.5	0.50	0.5	0.43
11. Paper making and manufactured goods	1.25	1.6	1.95	1.9	1.83
12. Print	)	1.0	0.95	0.9	0.97
13. Cultural, educational and sports articles	) 2.41	0.5	0.49	0.5	0.50
14. Arts and crafts	)	0.8	1.02	1.1	1.13
15. Power generation, steam and hot water production and supply	6.12	5.5	4.04	4.1	4.77
16. Petroleum processing	3.80	3.9	3.27	2.7	2.55
17. Coking, gas and coal-related products	0.12	0.2	0.15	0.2	0.20
18. Chemical industry	)	6.1	7.25	7.5	7.76
19. Medical and pharmaceutical products	) 11.21	1.3	1.86	1.7	1.84
20. Rubber manufactured goods	)	1.7	1.30	1.4	1.51
21. Plastics manufactured goods	)	1.4	1.71	1.6	1.63
22. Building materials and other non-metal mineral products	4.62	5.9	6.05	5.8	5.38
23. Smelting and pressing of ferrous metals	5.82	6.2	6.36	6.6	6.17
24. Smelting and pressing of non-ferrous metals	1.49	1.8	1.87	2.1	1.93
25. Metal products	)	2.7	2.74	2.8	2.77
26. Machine building	)	11.6	10.99	10.6	9.50
27. Transportation equipment	)	4.4	3.65	3.6	3.67
28. Electric equipment and machinery	) 21.89	4.0	4.10	4.5	4.12
29. Electronic and telecommunications equipment	)	2.8	2.94	2.9	2.87
30. Instruments, meters and other measuring equipment	))	1.1	0.89	0.9	0.79

Source: State Statistical Bureau of China.

Fig.III.1. Growth of net industrial output by major branches of industry, 1985 - 1990, (Percentage)

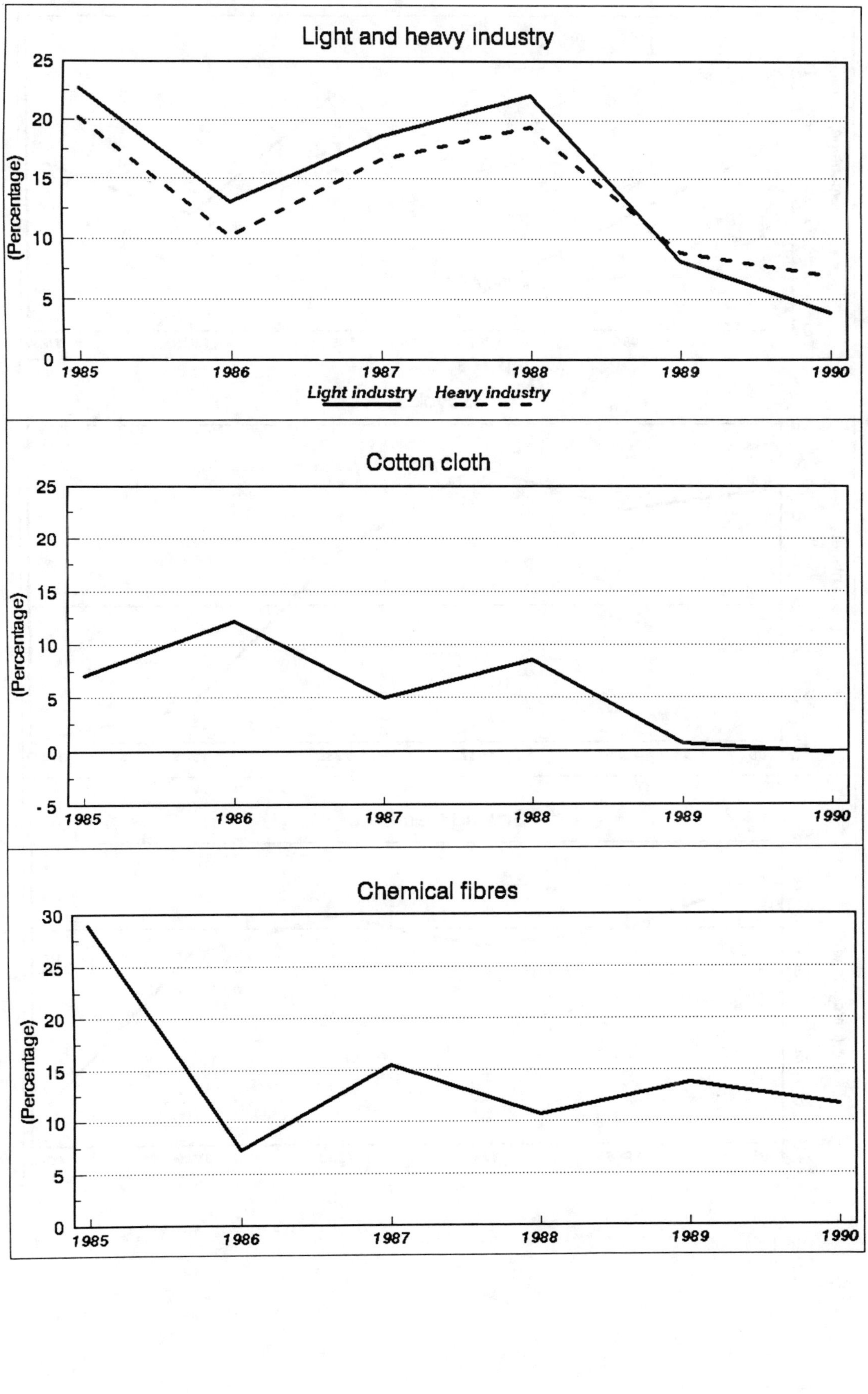

Fig.III.1. Growth of net industrial output by major branches of industry, 1985 - 1990 (Percentage)

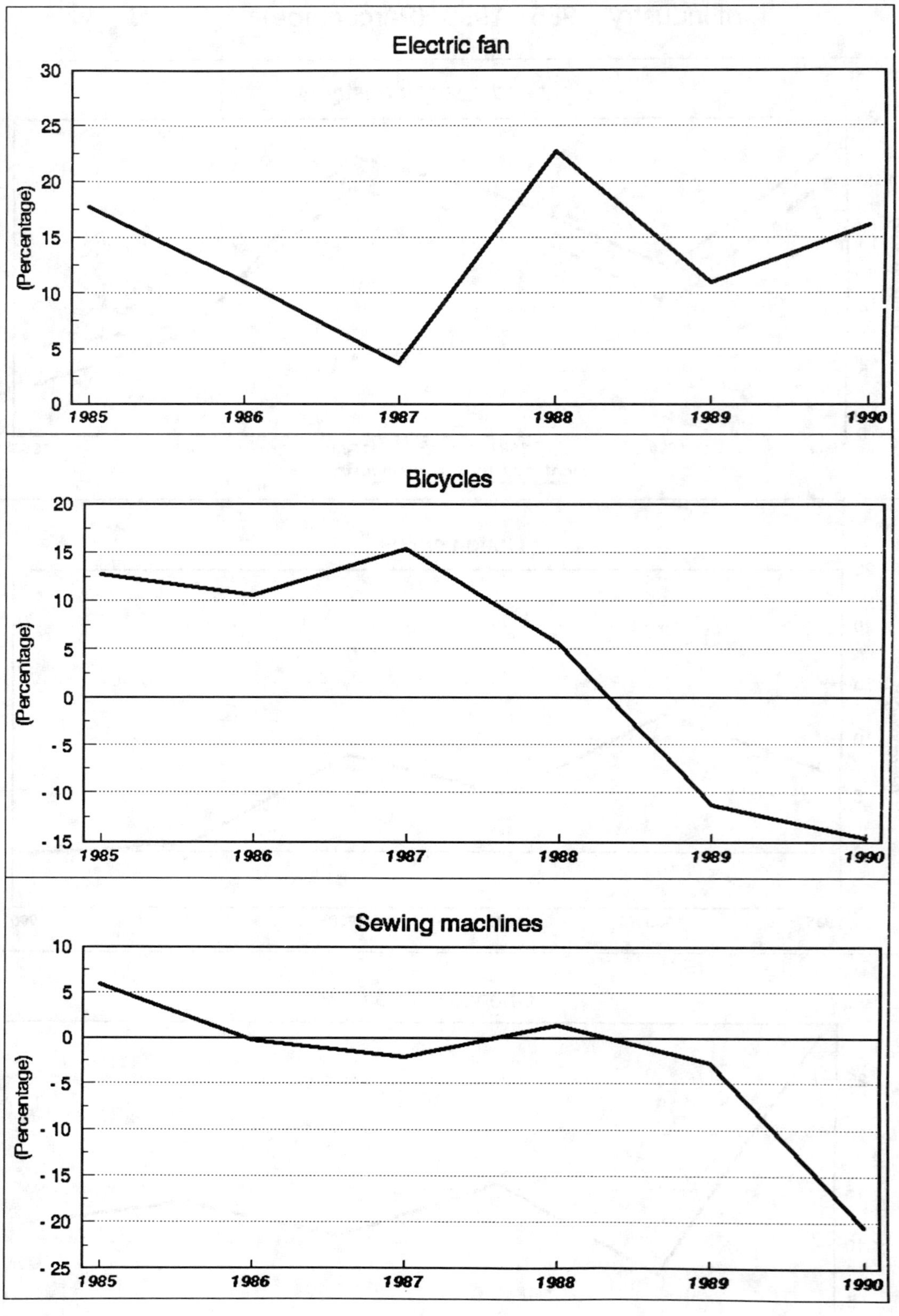

Fig.III.1. Growth of net industrial output by major branches of industry, 1985 - 1990 (Percentage)

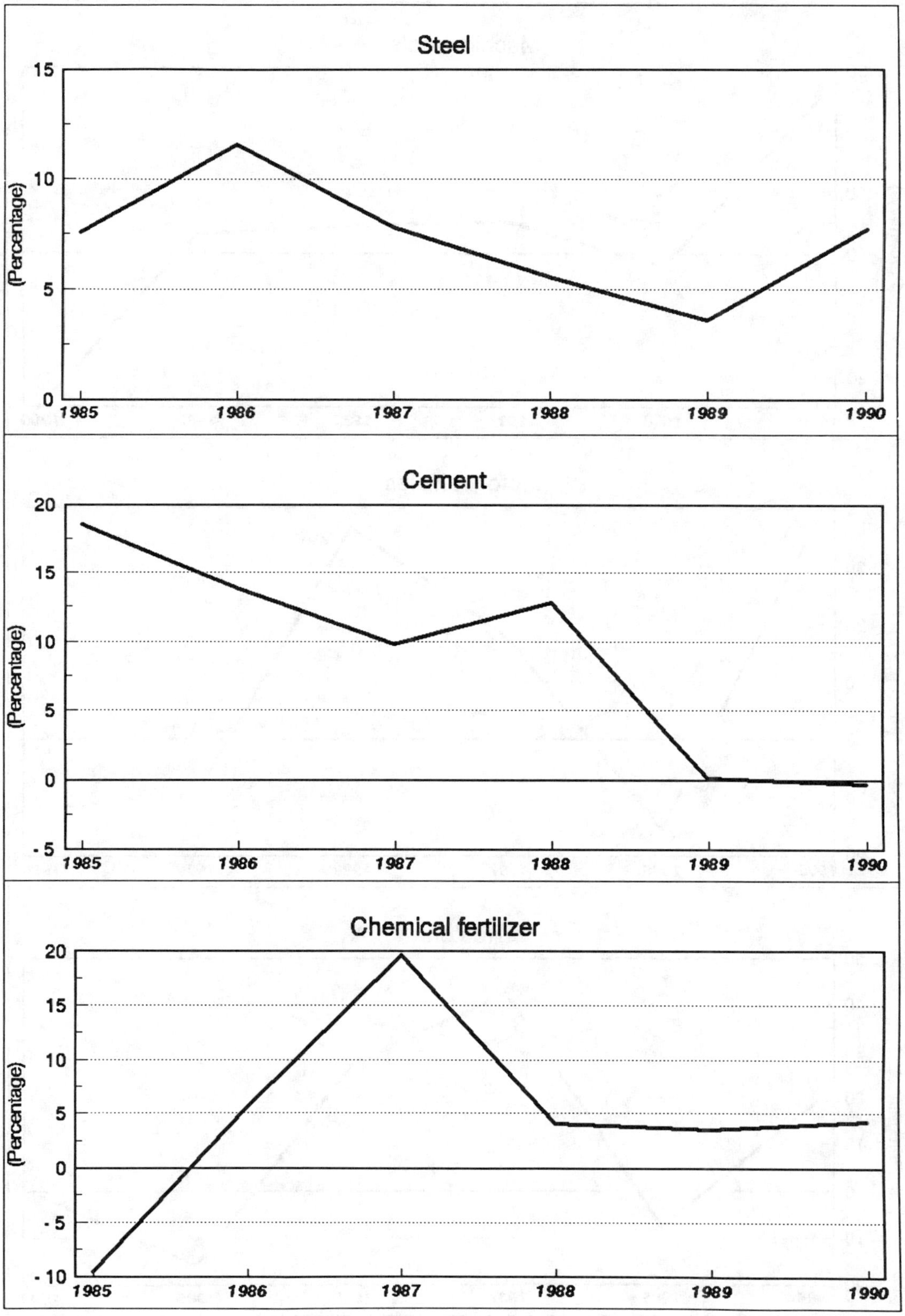

Fig.III.1. Growth of net industrial output by major branches of industry, 1985 - 1990 (Percentage)

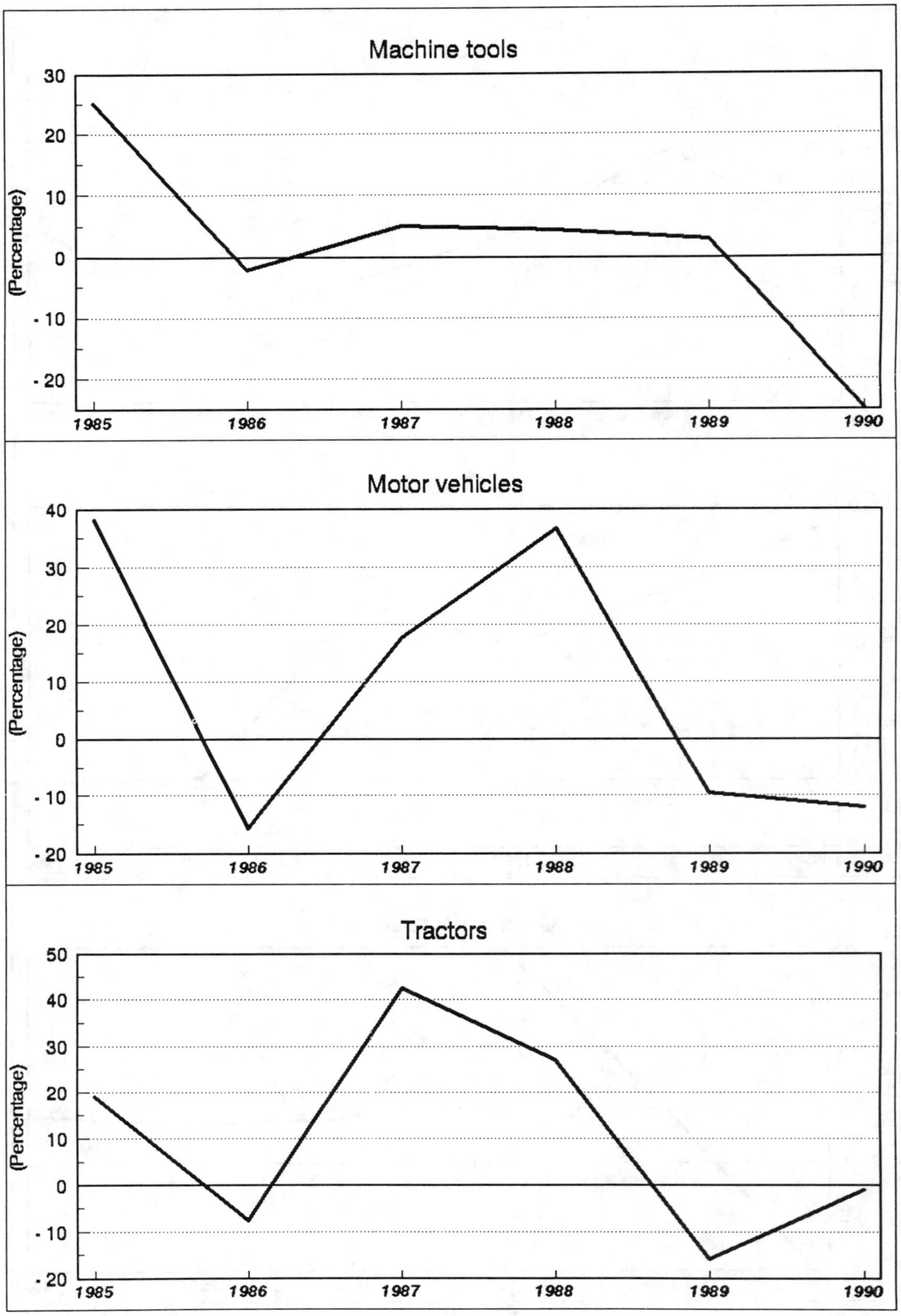

Fig.III.2. Structure of net industrial output, 1982 (Percentage)

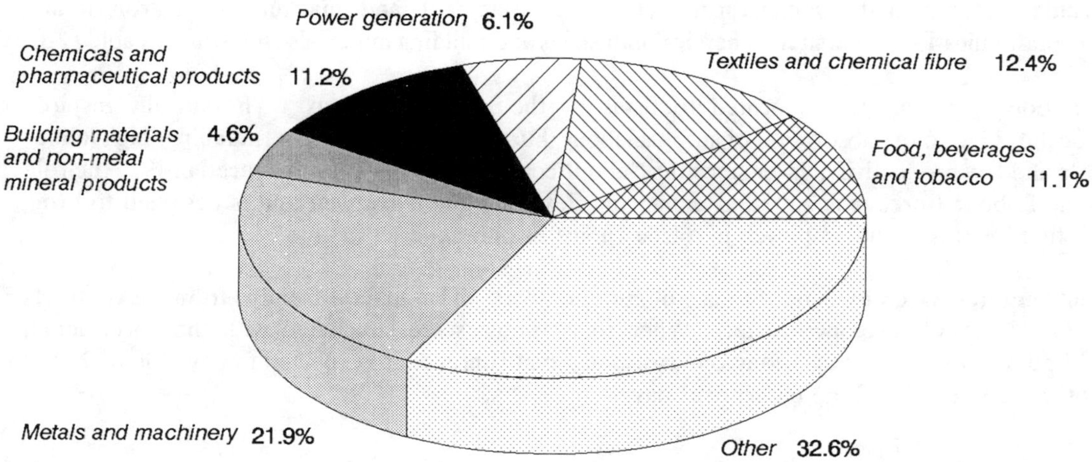

Fig.III.3. Structure of net industrial output, 1990 (Percentage)

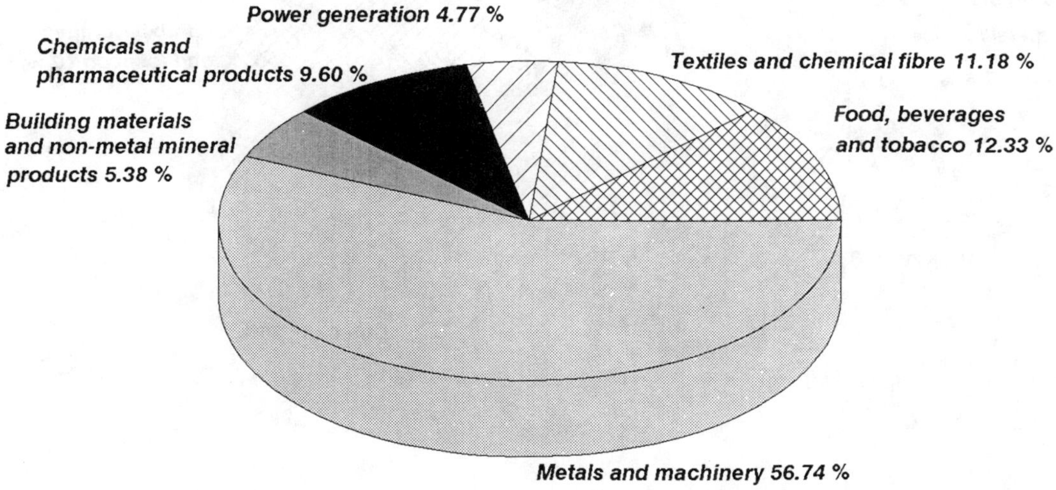

B. MANUFACTURING EMPLOYMENT, LABOUR PRODUCTIVITY, WAGES AND SALARIES

Towards a free labour market

The manufacturing sector provided employment for 121.16 million persons, representing 21.9 per cent of China's total labour-force of 553.2 million in 1989. Disaggregate data on manufacturing employment in State-owned enterprises,[3/] which account for around one-third of manufacturing employment, show that the machine-building subsector is the largest employer of the industrial labour force. Other subsectors making significant contributions to manufacturing employment include textiles, food manufacturing, electrical equipment and machinery, electronics and telecommunications equipment, chemical industries and building materials (see Annex Table A-8).

Traditionally labour was allocated to firms by the Labour Bureaux. The socially inspired allocations led to the absorption of surplus labour, burdening firms with overmanning, high labour costs and low productivity. Ever since the "labour contract system" was introduced in 1979 the role of the Labour Bureaux is aimed at enhancing flexibility and dynamism and is designed to bring together employer and employee in a free choice of option.

Joint ventures have pioneered the use of contract work. The general trend is to make contracts less binding and severance easier. When joint ventures are concluded with the government, minimum pay scales are agreed upon. Normally joint ventures are expected to pay at least 20 per cent more than what State enterprises pay.

Fig.III.4. Structure of manufacturing employment, 1990 (Percentage)

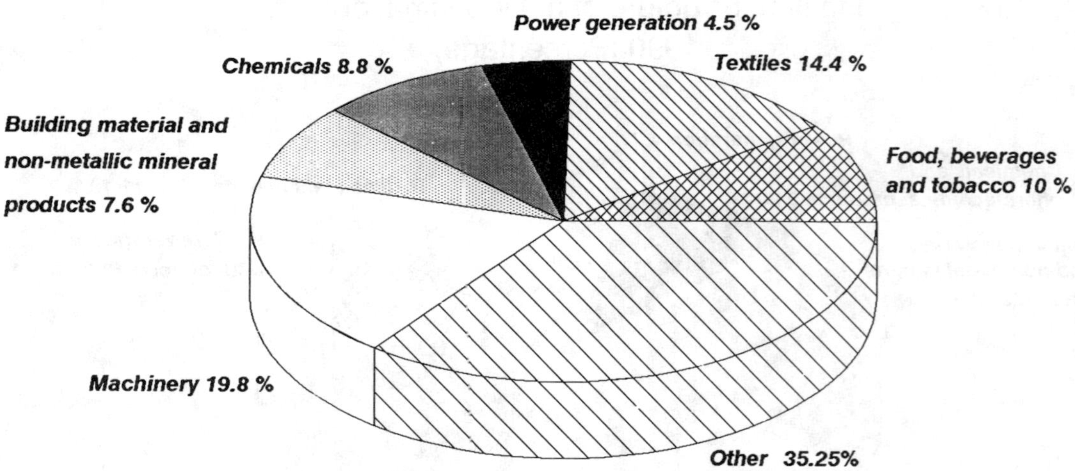

Efficiency-linked pay rise

Wage increases were virtually frozen during 1963-1977. By 1977 the average wage was lower than in 1952. Wage increases since 1977 have gone a long way towards meeting the 15-year backlog of demand for high wages. The average industrial wage per year (in enterprises run by foreign businessmen) has risen from Rmb 1,737 in 1984 to Rmb 3,411 in 1990 (see Table III.3), a nominal increase of 11.9 per cent per annum. State policy encouraged enterprises to link pay increases to skill and work efficiency. Incentive pay and the bonus system significantly altered the wage bill.

Table III.3. Average annual wage of staff and workers in enterprises by type of ownership, 1984-1990
(Rmb)

Type of ownership	1984	1985	1986	1987	1988	1989	1990
State-collective joint ownership	949	1,172	1,358	1,527	1,943	2,184	2,295
State-private joint ownership	1,835	2,194	2,213	2,406	3,229	3,383	3,780
Collective-private joint ownership	932	1,612	1,636	1,723	2,255	3,051	3,144
Joint ventures with foreigners	1,655	2,111	2,082	2,245	2,447	2,669	2,905
Enterprises run by overseas Chinese and Hong Kong and Macao Chinese	1,113	2,500	1,613	1,830	2,966	2,995	3,687
Enterprises run by foreign businessmen	1,737	2,144	2,380	2,826	2,012	3,567	3,411
Other	1,303	1,247	1,245	2,160	2,274	2,763	2,448

Source: State Statistical Bureau of China, *China Statistical Yearbook* (various issues).

Wage reforms brought about some improvements in labour productivity. Empirical studies reported a slow increase in labour productivity growth until the early 1980s. In certain key industries, such as electrical power, coal, petroleum and lumber, labour productivity remained unchanged or declined. This does not reflect worker performance alone. Transport problems, raw material shortages, energy bottle-necks and inefficient means of production also affected output per worker.

Overall labour productivity data for 1987 and 1988 depict a strikingly different picture (see Table III.4). Labour productivity in industrial enterprises with independent accounting systems rose significantly in real terms across several subsectors of manufacturing. The growth of labour productivity, measured in constant 1980 prices, exceeded 40 per cent in consumer electronic apparatus and consumer electronic equipment. Over 20 per cent growth in labour productivity was recorded in electric equipment and machinery, transport equipment, plastic products, arts and crafts articles, furniture manufacturing, and the manufacture of animal fodder.

In 1988, industrial wages rose by 28.1 per cent in nominal terms. Adjusting for the official inflation rate of 18.5 per cent, wages rose by 9.6 per cent in real terms. Thus the growth of labour productivity in the above segments of manufacturing was much faster than increase in real wages in 1988. With the exceptions of cotton textiles, power production and supply, petroleum processing and smelting and pressing of non-ferrous metals, the growth of labour productivity was significant in almost all subsectors of manufacturing. In 1990, cotton textiles and power production and supply recorded improvements in labour productivity.

China has invested heavily in industrial modernization in the 1980s with a view to achieving productivity gains through technical change, scale economies, a more efficient allocation of resources or through a combination of these three sources. Recent findings[4/] based on data from 293 Chinese industries show that substantial productivity gains can be achieved by transferring technology from the State to the collective sector and private enterprises. Re-allocating investment and labour could significantly enhance productivity.

Table III.4. Overall labour productivity by industrial subsectors with independent accounting systems, 1987-1990
(Rmb/person-year, at constant 1980 prices)

Industrial branch	1987	1988	1989	1990
Food manufacturing	20,167	22,130	22,506	23,586
Grains processing	34,900	37,234	36,838	36,676
Beverage manufacturing	14,013	15,974	18,438	17,212
Tobacco manufacturing	81,589	90,483	92,797	99,218
Fodder manufacturing	40,813	51,017	49,087	47,900
Textile industry	16,417	17,223	17,267	17,857
Cotton textiles	18,199	18,559	18,303	18,656
Woollen textiles	17,209	18,707	17,550	17,428
Silk textiles	13,195	13,975	14,809	16,759
Leather, fur and their products	10,741	12,175	13,266	15,311
Timber processing, bamboo, cane, palm fibre and straw products	7,253	8,321	8,333	8,212
Furniture manufacturing	8,702	10,814	11,786	11,944
Paper making and paper products	13,322	14,758	15,839	16,559
Printing	10,051	11,519	12,487	13,416
Cultural, educational and sports articles	12,162	13,491	14,399	15,512
Arts and crafts articles	8,111	10,214	11,910	12,985
Power generation, steam and hot water production and supply	23,775	24,293	24,877	25,015
Petroleum processing	73,862	74,541	77,866	77,317
Coking gas and coal related products	11,369	11,786	11,798	13,365
Chemical and allied products	18,706	20,589	21,331	22,669
Basic raw chemical materials	13,949	15,254	15,929	16,510
Organic chemical products	27,398	30,030	30,978	33,159
Consumer chemical products	26,090	29,159	29,650	33,700
Medical and pharmaceutical products	30,029	34,743	33,897	37,961
Chemical fibres	43,234	48,373	53,124	59,808
Rubber products	21,200	23,161	25,111	26,129
Plastic products	15,309	18,963	19,644	20,979
Building materials and other non-metal minerals	6,010	6,864	7,268	7,916
Cement manufacturing	6,623	7,334	7,171	7,683
Smelting and pressing of ferrous metals	19,160	19,607	19,382	20,475
Smelting and pressing of non-ferrous metals	25,949	25,759	26,966	26,709
Metal products	11,082	12,693	14,021	15,009
Consumer metal products	12,570	14,418	14,776	16,508
Machine building industry	12,001	14,136	14,629	14,522
Industrial machinery	11,491	13,680	14,389	14,264
Consumer machinery	22,046	23,279	22,888	22,575
Transportation equipment	13,368	16,734	18,104	18,797
Electric equipment and machinery	19,071	22,897	25,050	24,899
Consumer electric equipment	33,252	47,523	51,554	50,181
Electronic and telecommunications equipment	28,604	38,235	38,193	41,132
Consumer electronic appliances	60,865	86,177	78,485	75,293
Instruments, metres and other measuring equipment	10,867	12,949	13,571	13,485

Source: State Statistical Bureau of China.

As the Chinese industrial sector is locked into the economic rectification programme, job seekers find it increasingly difficult to find jobs in the readjustment of industrial structure. By end-1989, more than 3.78 million persons were seeking jobs[5] in urban areas. There are indications that the number of job seekers will continue to rise in 1991. According to official statistics, 80 per cent of the State-owned enterprises are 15 to 20 per cent overstaffed. In other words, there are altogether 15-20 million persons in need of reassignment. This calls for prudent manpower planning and human resource development.

Human resource development[6]

Human resource development in China is characterized by two basic tenets: population growth control and full employment. China's population totalled 1,143.3 million as of December 1990, which represents over 20 per cent of the total world population. As a result of the strict family planning policy adopted since 1973, the annual population growth rate declined from 2.3 per cent to 1.3 per cent in the 1980s. Nevertheless, the target of limiting the population to 1.2 billion by the turn of the century seems difficult to achieve given the recent upward swing in the birth rate and the population may reach 2 billion by the 2030s. In order to provide such a vast population with a sufficient number of jobs, sustainable economic growth has to be concomitantly accompanied by an upgrading of the quality of human resources.

An important aspect of China's human resource development is related to educational development from basic education to professional training. When the People's Republic of China was founded in 1949, 80 per cent of a population of 450 million was illiterate. During the period of economic recovery from 1950 to 1958 a new education system was established. In accordance with the First Five-Year Economic and Social Plan (1954-1958) basic education and higher education system was structured, and elimination of illiteracy was an urgent task in its national planning.

As a result, despite having experienced a decade of political turmoil which started in 1966, 1,500 million people are estimated to have received a basic education in the past three and half decades. China's literacy rate has risen to almost 80 per cent and the enrolment rates in primary education for the age group between 6-7 and 12-13 reached 96 per cent in 1986. However, China's higher education has remained at a low standard and there is a minimal enrolment rate for the vast population. During the Cultural Revolution period, in particular, the whole system of intellectual education was disrupted and consequently universities failed to produce quality graduates.

Indeed the quality of labour is a major obstacle for China's industrial development. Data for the early 1980s indicate that only less than one per cent of the total workforce received college or higher level education, 10 per cent had high school education, and the remaining 89 per cent had an education of junior high school or less, including 28 per cent of illiterate workers.

Under the past education policy the system produced only a small academic elite who tended to over-specialize in narrow fields and also to specialize in fields of studies irrelevant to the productive sectors. Since China undertook the policy of reform in 1978, however, the education system has been reestablished and adjusted in accordance with social and economic needs. According to the policy reform, science and technology have been officially seen as the motors of the future growth and modernization of China's economy. The government has endeavoured to formulate and implement the policy to reform the science and technology sector with the assurance that it will be strongly linked with social and economic development. It was in this context that policy-makers started to focus their attention on the higher education sector. Today it is well recognized by the government and people in China that upgrading intellectual manpower is a basic requirement for the modernization drive and economic and social development.

To meet the demand for skilled labour in industrial development it is a prerequisite to establish vocational and technical training institutions both in urban and rural areas in China. The policy of "training before employing" has been actively promoted, and employees in all sectors of the economy have been recommended to receive on-the-job as well as part-time training opportunities to upgrade the quality of labour.

According to a forecast, China's agricultural labour force will drop from 73 per cent of the total to some 54 per cent by the year 2000, while rural industrial employment will rise from 34 million to 150 million. This implies that more than 100 million of the rural industrial labour force will need training to meet their job requirements. To this end vocational and technical schools and other technical training facilities will play a significant role in producing a skilled industrial labour force in rural areas.

Having realized all the problems of personnel management derived from the model of the centrally planned economy, the government has attempted reforms many times in the past to resolve the problems. However, the reforms were restricted only to adjusting the administrative power of central and local authorities and of the different departments and regions, the measures of giving autonomous managerial authority to industries and enterprises were not adopted at a full scale in the national restructuring programmes. Reforming of the whole structure of the human resources management system constitutes one of the critical aspects of China's human resource development.

One problematic question is whether the emphasis on rapid growth and material incentives could lead to the possible emergence of structural inequities among the sexes: a small elite group of men with high incomes earned through private economic activity and a large underclass of unskilled female workers. The status of women in traditional China might still unconsciously influence the image of women in modern China. The ideology regarding sex roles is an eminent factor in explaining why women's work takes a specific form and is given particular status.

The economic and social status of women in general could be improved through selective support in literacy and skill training courses carried out collaboratively with the All China Women's Federation. However, for being competitive also the women entrepreneurs need training. The All China Women's Federation and UNIDO agreed to develop a training programme for women entrepreneurs to improve their management and business skills. The development of a training programme is in line with the changing demands in the Chinese economy for women entrepreneurs. Such a training programme fits in the context of the Third UNDP Country Programme which places emphasis on "increasing efforts to improve the status of women in poor rural areas".

C. PERFORMANCE AND EFFICIENCY

Market forces to clear distortions

There had been a marked improvement in investment efficiency during 1978-1986. Total factor productivity growth, which had been negative during 1957-1965 and zero during 1965-1977, increased to 3.8 per cent per annum during 1978-1986. However, a close look at the performance and efficiency of State-owned large-scale heavy industries and other categories of enterprises reveals mixed trends.

A striking feature of industrial expansion since 1984 has been an increased reorientation towards the market economy, as enterprises altered investment and production patterns in order to respond effectively towards rapidly changing demand patterns. Self-financing became an important source of investment for all categories of enterprises. It was estimated that during the second phase of the reform period (1984-1988) even State enterprises financed 20 per cent of fixed capital formation and about a third of modernization through retained profits.[7/]

The increasing importance of the market is reflected in the fact that by 1988 the output value of products subject to mandatory planning accounted for about 20 per cent of total industrial output value, compared with 40 per cent in 1985. By 1986, 76 light industrial products were subject to the guidance plan, and only salt, cigarettes, newsprint and anastatic printing paper came under the mandatory plan. All other light industrial products were regulated through the market. As a result, commercial enterprises have been free to draw up their own purchasing plans and contracts, and been able to select the most suitable time to buy their materials. This has led companies to negotiate their own sales. In 1987, two-thirds of the output of 24 major products were sold by commercial enterprises themselves.

With the reform of the planning and material supply systems, the number of materials distributed solely by the State fell from 279 in 1979 to 24 in 1987. Likewise, the proportion of materials distributed according to a central plan has also dropped. In 1988, the State distributed 46.8 per cent of steel products, 43.5 per cent of coal, 13.6 per cent of cement and 25.9 per cent of timber.

The proportion of materials subject to unified distribution by trade has also fallen. During the Fifth Five-Year Plan (1976-1980), 81.6 per cent of steel products were centrally distributed by the State machine-building industry. In the Sixth Five-Year Plan (1981-1985), it was 52.4 per cent. And in the first three years of the Seventh Five-Year Plan (1986-1988), the figures have been 45.8 per cent, 33.6 per cent and 28.4 per cent, respectively. With the introduction of market mechanisms, enterprises have been transformed from applying pure production functions to applying basic production and management principles. A sample survey revealed that 80 per cent of companies believed that market changes should be taken into account in planning production, and 90 per cent believed raw materials should be purchased on the market.[8/]

All the companies surveyed in the study sold their products through their own channels and said this had encouraged them to cut down on wastage of materials and improve quality in order to enhance economic return. In the past ten years, energy consumption for every Rmb 100 million worth of output values has fallen by 30 per cent in Chinese enterprises and the amount of steel products per unit of production consumed has dropped by 23 per cent.

Economies of scale

Efficiency levels have also improved since the adoption of measures by the State Council to rationalize production structures and to construct industrial conglomerates, with a view to reducing production duplication and to benefiting from economies of scale. This has led to an increase in the level of industrial concentration and to the emergence of about 20 giant corporations which handle key industrial sectors and which are in a position to influence their price and output-level decisions on a wide range of markets.

The degree of capital intensity

Table III.5 compares the gross and net value of industrial output of different types of enterprises. It is evident that the share of net output in total industrial value in China is only moderately higher than in middle-income developing countries.[9/] The net to gross output ratio is highest for State-owned industry, heavy industry and large-scale units. This reflects both the relatively higher capital intensity of State, heavy and large-scale production and also the fact that the prices of inputs of these enterprises are mandated and less subject to price fluctuations than small- or medium-sized firms.

Principal financial indicators

Some principal financial indicators of State-owned independent accounting enterprises are summarized in Annex Table A-9. These data reveal that the financial position of these firms had deteriorated significantly in 1990 relative to 1982. Although total funds employed in 1990 were in nominal terms more than double that in 1982, surplus (profits and taxes) earned had risen by only about 54.6 percentage points. The ratio of surplus to total funds had declined from 23.4 per cent in 1982 to 12.17 per cent in 1990. Both net output and surplus as a proportion of the value of fixed assets declined during 1982-1990.

Table III.5. Gross and net value of industrial output, 1989 and 1990
(Rmb 1,000,000 at current prices)

Type	Gross value of output		Net value of output		Net value as percentage of gross value added	
	1989	1990	1989	1990	1989	1990
State-owned	1,187,296	1,257,045	346,023	356,870	29.1	28.4
Collective	494,533	524,650	128,810	132,338	26.0	25.2
Other	65,560	87,227	15,508	20,117	23.7	23.1
Light industry	814,931	877,676	206,809	221,250	25.4	25.2
Heavy industry[a/]	932,458	991,247	283,532	288,075	30.4	29.1
Large-scale industries	476,930	650,921	183,208	199,998	31.8	30.7
Medium-scale industries	345,429	369,393	93,031	96,238	26.9	26.1
Small-scale industries	825,031	848,608	214,103	213,088	26.0	25.1

Source: State Statistical Bureau of China, *China Statistical Yearbook* (various issues).

a/ Including mining and logging.

A quick glance at net output and surplus generated by subsectors of manufacturing (see Table III.6) reveals that the surplus to net assets ratio is significantly higher for light industry relative to heavy industry. Taxes levied on light industry absorb a significantly higher proportion of total surplus generated. Despite this, light industry generally invests a larger proportion of its net profits than does heavy industry. If the highly taxed tobacco industry is excluded, the variation in average surplus (profit and tax) to net assets ratio across manufacturing subsectors however varies moderately. The most surplus-generating industries are petroleum refining, rubber manufactures, electrical equipment, metal products, beverages, chemicals and textiles. Branches with the lowest value of surplus to net value ratio for the 1985-1990 period are coal products, leather goods, food manufactures, timber products, non-ferrous metals and power generation, steam production, etc.

There is much scepticism about "profiteering" through the resale of raw materials and scarce commodities. Revenue from the sales of products of State-owned enterprises rose by 275 percentage points during 1982-1989. The proliferation of "profiteering" and other speculative activities led to distortions of the market mechanism. An investigation into the financial performance of State-owned enterprises uncovered Rmb 1.13 billion of illegal gains in 1989.[10/] Due to the public discontent, the government initiated several rounds of clean-up and rectification of profiteering activities. By June 1989, 6,481 State-owned enterprises were dissolved or merged, and 10,386 separated from their founding Party and government departments. In cleaning up and rectifying companies the government endeavours to create a better environment and conditions for reform and openness.

Amidst corrective steps there is a renewed emphasis on State enterprises. The profit-making non-State sector, which has grown from virtual non-existence before 1979 to about one-third of the national production in 1989 is seriously affected by the austerity programme. Credit for these enterprises have been squeezed. Non-State companies often outbid the State sector for raw materials. They can pay higher prices and pass them on to consumers by raising end-product prices. There are indications that the State sector will be sheltered and independent profit-oriented business will not be allowed to compete with State enterprises for scarce raw materials and energy.

Table III.6. Net output and surplus (profit and taxes) of national independent accounting units, by industry branch, 1985, 1988, 1989 and 1990

Branch of industry	Net output[a/]				Surplus[b/]			
	1985	1988	1989	1990	1985	1988	1989	1990
1. Food manufacture	40.11	41.10	39.70	35.97	22.71	20.79	16.97	11.42
2. Beverage manufacture	61.16	54.70	47.37	45.36	41.03	34.78	26.65	24.25
3. Tobacco manufacture	522.89	417.20	368.71	312.80	497.72	394.08	356.20	294.68
4. Forage manufacture	49.46	59.54	48.48	39.62	32.56	33.57	21.84	13.84
5. Textile manufacture	60.63	55.01	52.55	46.08	35.79	26.18	21.07	13.49
6. Chemical fibres	30.64	22.89	30.61	34.07	25.28	21.64	20.60	21.24
7. Clothing	118.61	92.07	95.06	87.24	49.45	32.03	28.83	23.64
8. Leather, furs and manufactured goods	72.46	65.57	61.42	57.88	34.84	23.54	14.73	9.10
9. Timber processing, bamboo, cane, palm fibre and straw products	44.64	36.66	40.57	30.91	21.66	19.34	12.60	3.97
10. Furniture manufacture	67.57	55.15	59.92	52.92	26.06	21.62	15.30	8.42
11. Paper making and manufactured goods	43.95	36.89	46.29	39.74	27.20	27.90	22.26	13.37
12. Printing	52.39	40.70	48.75	45.68	29.48	23.04	21.41	18.07
13. Cultural, educational and sports articles	97.57	77.79	77.22	69.71	60.53	40.51	31.20	24.33
14. Arts and crafts	117.97	86.02	101.33	91.79	49.24	37.62	30.13	24.81
15. Power generation, steam and hot water production and supply	17.76	12.42	13.27	13.68	12.50	10.32	10.39	9.93
16. Petroleum processing	89.47	53.13	44.30	36.26	79.66	47.95	34.08	27.28
17. Coking, gas and coal-related products	19.49	10.66	10.22	10.95	12.42	4.53	2.92	1.70
18. Chemical industry	30.92	31.21	40.90	37.08	20.70	27.29	23.22	17.99
19. Medical and pharmaceutical products	65.54	58.38	68.77	63.45	42.94	48.46	33.71	24.79
20. Rubber manufactured goods	95.37	60.93	73.84	72.37	69.05	42.72	40.36	31.62
21. Plastics manufactured goods	55.89	39.18	50.26	43.11	29.25	26.09	18.68	11.67
22. Building materials and other non-metal mineral products	41.10	32.81	38.23	32.77	20.67	17.87	14.45	9.42
23. Smelting and pressing of ferrous metals	30.54	26.27	33.00	28.35	23.84	22.56	21.55	16.51
24. Smelting and pressing of non-ferrous metals	30.20	24.38	34.82	29.49	18.29	19.59	19.13	13.08
25. Metal products	66.89	54.94	65.14	58.80	35.88	27.48	24.83	17.47
26. Machine building	39.57	35.69	43.76	37.44	22.13	19.64	16.35	9.32
27. Transportation equipment	39.86	29.31	38.70	37.59	24.53	18.60	15.22	11.45
28. Electric equipment and machinery	82.52	53.09	74.21	62.54	52.85	36.88	33.34	21.42
29. Electronic and telecommunications equipment	59.74	45.31	64.06	55.09	36.11	33.58	24.59	16.03
30. Instruments, meters and other measuring equipment	56.03	42.38	53.44	45.15	32.10	21.88	19.39	13.05

Source: State Statistical Bureau of China.

a/ Net output value per Rmb 100 original value of fixed assets.
b/ Profits and taxes per Rmb 100 original value of fixed assets.

Much of the industrial inefficiency currently stems from China's inability to produce raw materials and energy in sufficient quantities. Although non-State enterprises generate a much higher rate of surplus, State enterprises benefit from low input prices. Unlike non-State enterprises the State-owned firms are less likely to be able to pass on cost increases in the form of higher output prices. As long as price structures do not reflect scarcity values it will be extremely difficult to gauge efficiency levels and growth within Chinese industry.

D. MANUFACTURED EXPORTS AND IMPORTS

Accelerating the export drive

Manufactured exports in total exports rose from 50.0 per cent in 1979 to almost 74.4 per cent in 1990. The largest single commodity group within manufactured exports are light industrial products. As against the depressed local demand, there has been a significant improvement on the export front in 1989. Buoyed by the strong export momentum in 1990 amidst a drastic cut in imports, China recorded an overall trade surplus in 1990.

Comparative advantage of exports

Export diversification had not been particularly rapid during the 1970s or 1980s. Textiles and food products continue to dominate China's export profile. The overwhelming evidence from the statistical estimates prepared by UNIDO for the early 1980s revealed that China has increased its comparative advantage[11/] in natural resource-based industrial products, lost some ground in labour-intensive products (particularly textiles) and failed to make any significant headway in technology- and capital-intensive export markets. The country's comparative advantage seems to be in products which have experienced declining shares in world trade. China has the lowest comparative advantage in capital- and technology-intensive products which are rapidly increasing their share in world trade.

Fig.III.5. Share of manufactured exports in total exports, 1990 (Percentage)

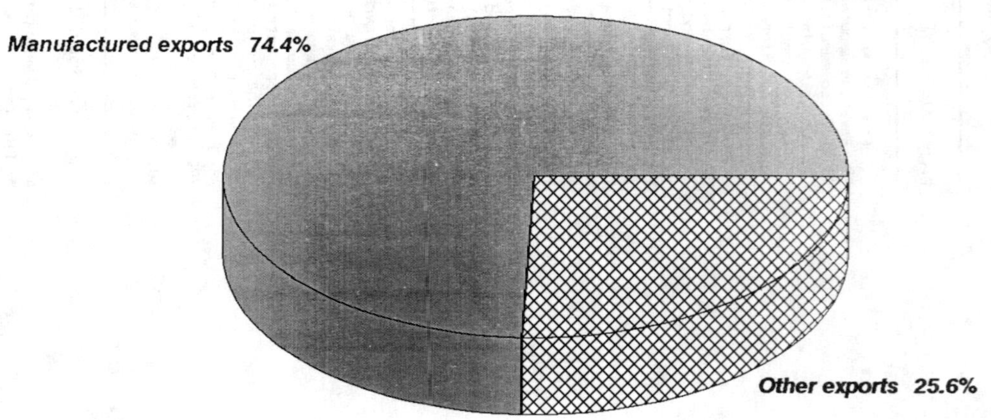

Export performance of selected areas

Nevertheless, manufactured exports continued to grow rapidly in recent years - particularly from the areas which have been established in order to attract foreign investment and in which State regulation of markets has been minimized. The total volume of exports of the four Special Economic Zones (SEZs) in 1987 was 19 times that of 1980; exports of the 14 open coastal cities in 1987 went up 79 per cent compared with 1983; and goods exported by the open economic areas (i.e., the delta areas) in 1987 were double that of 1984 in value. The SEZs and the economic and technological development areas have become more export-oriented. In 1987, 40 per cent of the products made in the SEZs were destined for exports.

The 14 open coastal cities are now able to earn more foreign exchange by selling their products on the international market. For example, Dalian City has upgraded many existing industrial enterprises and established economic and technical ties with other areas or enterprises. This has enabled the city to double its annual exports. Since Shanghai revamped its machine-building industry, the industry's annual exports have grown by 35 per cent. Nantong City, which updated its textile industry, reported that 65 per cent of its textiles were exported during the late 1980s.

Role of rural industry on the export front

Township and village enterprises (TVEs) have also become an important force in producing export commodities. The export goods supplied by rural township factories in Suzhou, Wuxi, Changzhou and Shanghai accounted for 20 per cent of all the exports of these four municipalities in 1987. While handicraft products and semi-processed goods were the major export items before, the rural township industries of these four municipalities on the Yangtze Delta are switching over to refined processing and medium- and high-grade products. Garments made in Qingpu County on Shanghai's outskirts, shirts and blouses tailored in Jiaxing County of Zhejiang Province, electric fans produced in Shunde County on the Pearl River Delta, lamps turned out in Zhongshan County and toys made in Dongguan County, also on the same delta, as well as hardware made in Dagang District of Tianjin City, have sold well abroad.

Major manufactured exports

China's major exports are listed in Table III.7. Products of the light industry dominate the country's export profile. In dollar terms, canned food, silk and satin material, fur shoes, sewing machines, bicycles and machine tools exports grew significantly in 1990. In the 1980s export growth has been stimulated by over 100 per cent depreciation of China's effective exchange rate.

China's exports of some specific manufactures are substantial in relation to world trade. The country's exports of silk, for example, constitute three-quarters of the global trade in silk. In some other categories, China's share in the world market is small, but is tending to grow rapidly. These products include textiles, clothing, footwear, floor coverings, fur skins, explosives, ceramic and glassware products.

Like other saleable commodities, silk exports used to be a monopoly of the Ministry of Foreign Relations and Trade (MOFERT) in Beijing. But such a centralized system proved to be both inefficient and stifling. To stimulate exports, the government began to experiment with various measures aimed at reducing the trade monopoly of MOFERT's import and export corporations while giving more freedom to provincial authorities and individual silk producers to trade on their own account.

Major imports

Around 80 per cent of China's imports comprise raw materials, industrial machinery and transport equipment. In an attempt to save foreign exchange earnings, imports to China are being reduced drastically. However, import restrictions are taking their toll on industrial expansion and modernization. A 39-point policy document adopted by the Party's Central Committee in

Table III.7. Major manufactured exports, by volume and value, 1982, 1985-1990

		Volume							Value(in $ million)						
		1982	1985	1986	1987	1988	1989	1990	1982	1985	1986	1987	1988	1989	1990
Canned fruit	('000 tons)	59.5	40.7	47.9	87.3	88.0	71.4	77.8	38.5	27.5	35.2	59.9	58.7	51.7	53.3
Beer	('000 tons)	26.0	28.0	28.4	32.4	39.3	41.8	35.2	7.4	9.2	10.9	13.6	20.0	25.2	19.6
Cotton yarn	('000 tons)	48.7	154.8	228.3	242.8	205.7	183.7	176.2	118.6	294.4	424.5	535.1	511.8	424.2	390.2
Cotton cloth	(million metres)	1179.0	1573.4	2056.3	2341.5	2229.5	2338.3	2221.6	717.5	994.3	1269.7	1525.3	1487.3	1600.5	1602.4
Filature silks	('000 tons)	10.4	10.9	9.4	9.2	9.4	11.4	7.6	283.3	250.9	232.6	233.0	303.7	526.9	362.1
Silk and satin materials	(million metres)	72.9	114.7	161.2	150.4	209.9	173.1	173.8	189.7	312.5	426.1	459.0	679.6	716.1	769.8
Fabric woven	(million metres)	558.2	735.7	971.2	1147.4	1037.9	1147.0	1149.3	325.6	356.5	463.9	632.3	619.8	685.2	676.2
Tea	('000 tons)	105.8	136.8	172.1	174.3	198.4	204.6	195.5	224.8	294.1	327.0	362.5	401.9	420.7	412.7
Jute bags	(million units)	174.5	177.1	227.7	143.5	126.0	126.2	115.2	66.3	80.9	72.0	47.7	43.3	42.5	37.9
Bristles	('000 tons)	6.9	9.0	9.4	11.1	11.0	13.0	12.1	53.7	72.6	83.0	99.8	98.6	95.8	71.1
Fur shoes	(million pairs)	11.2	8.0	13.3	26.0	31.4	44.2	66.7	48.4	39.4	67.0	124.6	173.1	258.1	427.9
Carpets, superior quality handmade	(million sq m)	5.0	5.1	5.3	6.7	7.2	8.3	8.3	152.1	201.7	217.6	285.1	389.8	431.3	390.2
Goat skins	(million hides)	17.6	17.8	14.2	9.4	11.5	7.5	9.1	63.5	58.8	49.9	34.2	42.3	30.1	34.1
Fur mattresses	('000 pieces)	3062.9	2820.0	2740.0	2320.0	2020.0	1380.0	1110.0	41.3	26.3	25.0	29.2	27.6	17.4	12.5
Paper	('000 tons)	219.8	185.1	258.3	321.0	259.8	219.6	130.4	102.5	109.1	145.0	140.3	136.6	138.9	94.6
Sewing machines	('000 units)	460.0	330.0	600.0	590.0	800.0	1230.0	2160.0	22.0	11.0	16.7	20.2	28.1	44.0	70.9
Bicycles	('000 units)	610.0	540.0	410.0	820.0	1510.0	2450.0	3780.0	26.5	20.4	16.0	33.3	59.4	96.5	144.9
Tyres	('000 sets)	440.0	370.0	630.0	1029.0	1240.0	1200.0	1660.0	25.7	24.5	51.9	54.2	62.2	67.3	67.3
Machine tools	('000 pieces)	46.6	17.2	49.0	183.6	433.0	603.6	624.9	34.4	16.7	34.6	78.3	131.8	188.3	221.7

Source: Customs.

Fig.III.6. Value of selected manufactured exports, 1990 ($ million)

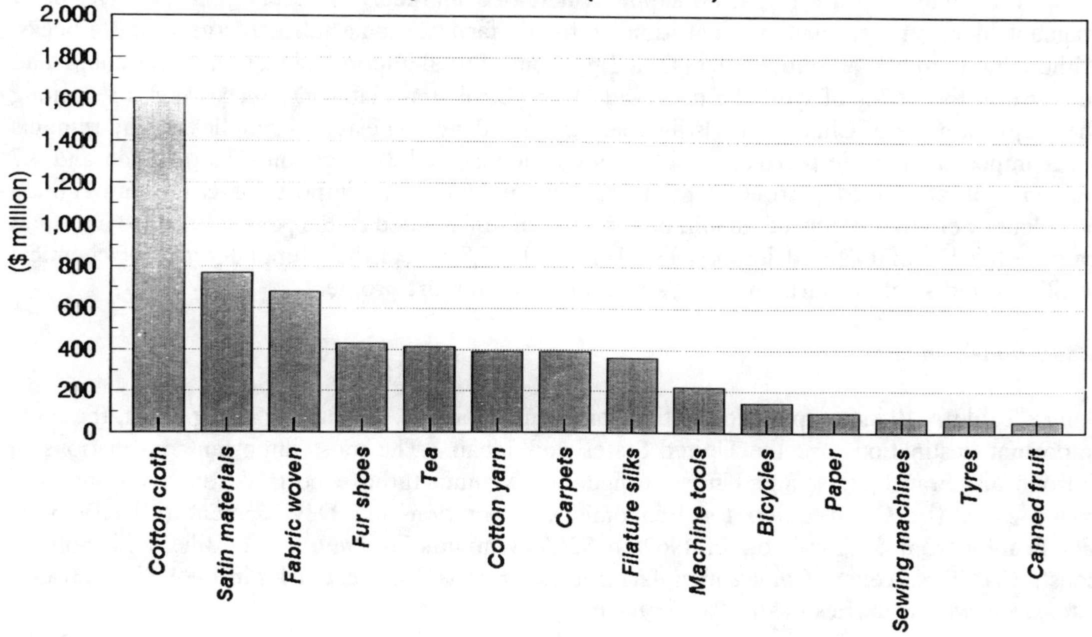

Fig.III.7. Destination of exports, 1990 (Percentage)

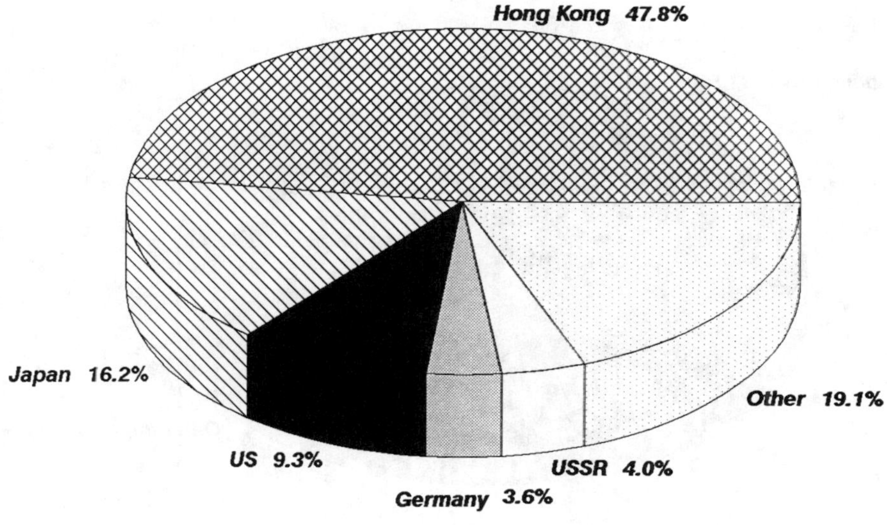

November 1989 described the 1984-1988 period as one of extravagance and over-spending and called for an absolute curb on imports of luxury goods and high quality consumer goods. The document also called for a sharp reduction in the imports of some unspecified raw materials.

China's dependence on external oil supplies increased markedly in recent years as a result of stagnant domestic production, inefficient refinery facilities and infrastructural bottle-necks. Although oil imports account for less than 3 per cent of total imports, higher oil prices could lead to a rise in the prices of synthetic fibre and other petroleum-related products which are among the major imports of China. Amidst increasing dependence on external supplies, China remains a net importer of crude petroleum. In 1989 China exported 24.4 million tons of crude and 4.7 million tons of refined petroleum products. In the same year imports of crude and refined petroleum were to the tune of 8.6 million tons, which represented an 8.6 per cent rise from a small base. Major manufactured imports listed in Table II.8 reveal the preponderence of chemical fertilizers and steel products in China's manufactured import profile.

Major trade partners

Annex Table A-10 lists export destinations and import origins. Excluding Hong Kong, the most important destinations are the United States and Japan. The most important destinations in Europe are France, Italy and United Kingdom. Manufactured exports destined for member countries of the Organization for Economic Co-operation and Development (OECD) rose significantly from \$257 million in 1969 to \$26,778 million in 1989.[12/] Textiles and clothing constituted 39 per cent of China's manufactured exports to OECD countries in 1989. China trades intensively with countries of the Pacific Basin.

In Europe, China is increasingly being exposed to severe competition from new entrants. In the regional market, the ASEAN countries' comparative advantages in similar product range tend to pose a potential threat to Chinese products.

Fig.III.8. Share of manufactured imports in total imports, 1990 (Percentage)

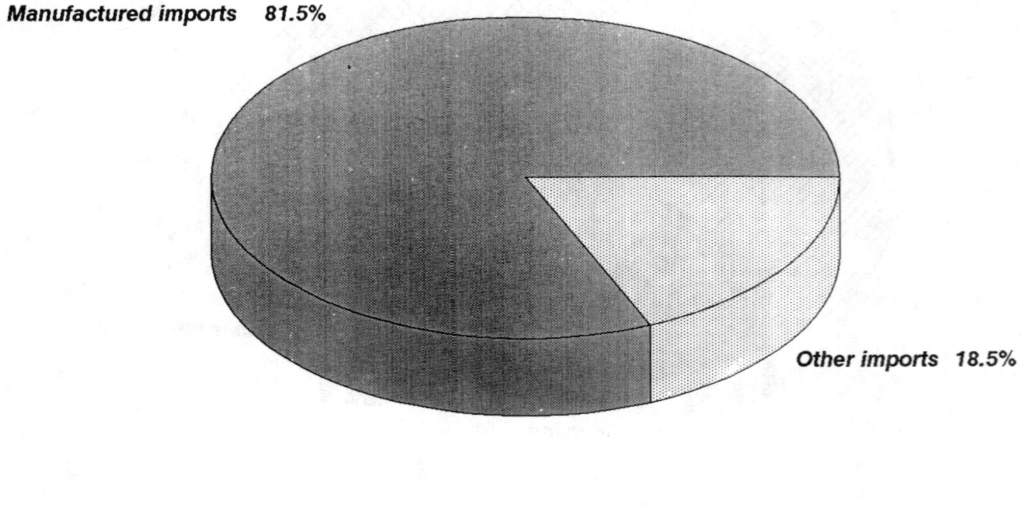

Fig.III.9. Value of selected manufactured imports, 1990 ($ million)

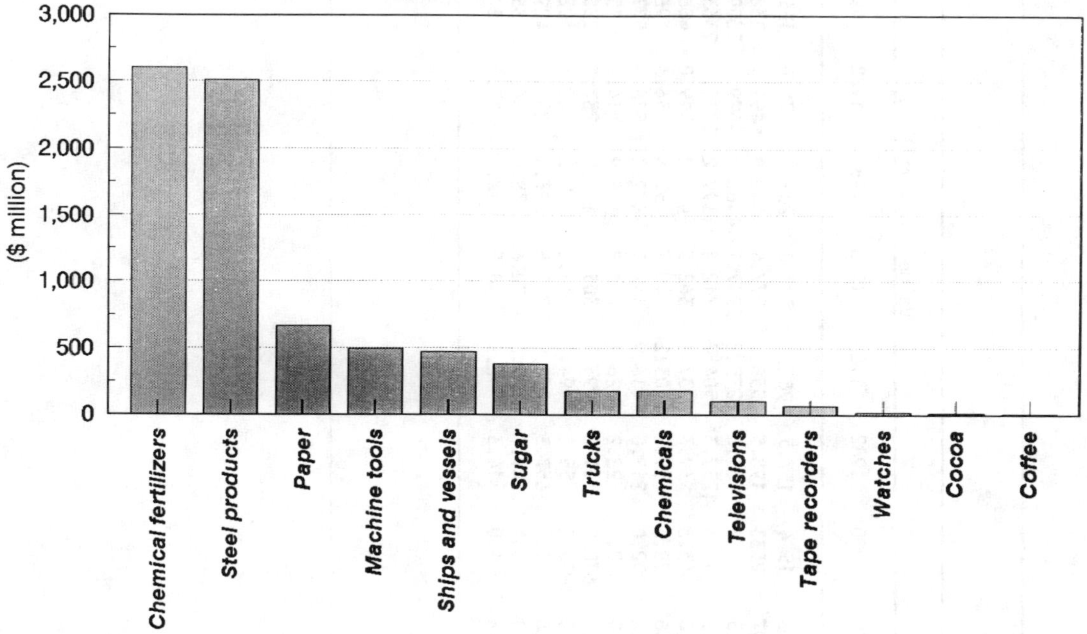

Fig.III.10. Origin of imports, 1990

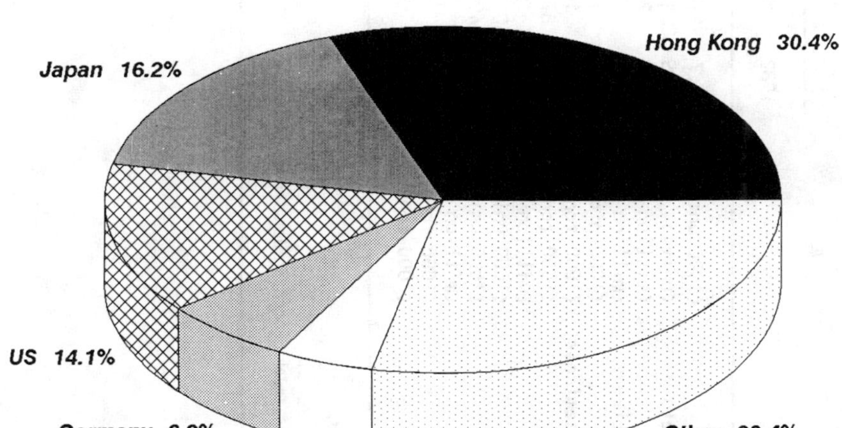

Table III.8. Major manufactured imports, by volume and value, 1982, 1985-1990

		Volume							Value (in $ million)						
		1982	1985	1986	1987	1988	1989	1990	1982	1985	1986	1987	1988	1989	1990
Trucks[a/]	('000 units)	17.7	111.5	64.6	19.2	14.2	12.6	15.4	157.5	982.7	807.0	205.9	171.7	125.9	171.7
Ships and vessels	(units)	295	1760	682	955	6066	11897	2533	198.5	558.5	277.9	148.8	191.5	338.0	468.2
Steel products	(million tons)	3.8	20.0	18.4	12.4	9.1	9.5	4.2	61.8	6275.2	5937.7	4462.0	4429.7	5600.6	2508.2
Chemical fertilizers	(million tons)	6.1	7.6	5.1	10.9	14.7	13.9	16.3	1252.3	1504.7	712.1	1399.2	2335.5	2363.7	2605.1
Machine tools	('000 pieces)	2.8	13.0	10.1	25.8	44.5	41.6	43.8	39.3	137.1	362.5	450.2	519.0	486.3	492.8
Agricultural chemicals	('000 tons)	15.9	16.1	7.5	10.1	34.1	36.6	28.5	59.2	103.2	40.7	55.8	156.3	196.2	177.1
Paper	('000 tons)	458.9	876.5	1095.3	1336.0	850.5	862.2	952.0	215.0	416.9	537.3	672.8	518.1	577.9	666.1
Watches	(million units)	1.5	5.2	2.9	1.8	1.0	1.5	2.8	22.2	26.1	13.8	3.9	14.1	15.7	16.8
Televisions	('000 units)	1035.0	5085.0	1391.7	1070.0	1286.8	1300.0	670.0	109.8	993.9	305.5	185.9	237.1	230.1	103.5
Tape recorders	('000 units)	1850.0	1496.1	522.5	2240.0	1675.7	4650.0	1930.0	55.1	94.2	34.0	49.2	40.6	102.3	63.4
Sugar	(million tons)	2.2	1.9	1.2	1.8	3.7	1.6	1.1	657.3	272.9	211.5	297.4	858.2	429.8	378.8
Cocoa	('000 tons)	22.7	6.3	27.2	14.5	16.8	24.0	10.1	47.7	15.5	72.6	38.6	34.9	35.5	11.9
Coffee	('000 tons)	14.3	0.5	1.8	1.6	2.8	6.2	1.0	14.3	2.9	8.5	12.9	17.5	17.9	6.3

Source: Customs.

a/ Includes chassis, trucks, jeeps, trailers, cabs, etc.

Most Favoured Nation status

In May 1990, the United States extended the Most Favoured Nation (MFN) status for China, which entitles China to the lowest available tariffs on products. Garments, toys, games and footwear account for more than half of China's exports to the United States. Had the United States revoked the MFN status, high tariffs would have been imposed on around 60 per cent of Chinese goods. On the other hand revoking MFN status would have forced China to impose higher duties on United States wheat, aircraft and aerospace equipment. There are indicators of extending MFN trade privileges for another year. As an act of reciprocity China decided in May 1991 to sign more than $1 billion worth imports from the United States. China's drastic fall in imports from the United States has been an issue in the MFN debate. It has become increasingly important for China to accept positive reciprocity in trade as China endeavours to resume its membership of General Agreement on Tariffs and Trade (GATT). With a view to making China's trade system more transparent and acceptable to other GATT members, it is necessary to make clear that the recent cutbacks in imports and slow-down in the pace of reforms are a short-term reaction to an overheated economy and do not represent a more permanent change of policy approach.

E. INVESTMENT AND OWNERSHIP PATTERNS

Investment in fixed assets

Until recently, rapidly rising levels of investment have been a salient characteristic of Chinese development experience. Between 1979 and 1987, China invested Rmb 1,644.1 billion in fixed assets. Of this total, Rmb 1,105.2 billion was invested in State-owned enterprises, Rmb 340.8 billion more than in the preceding 26 years (1954-1978). China has completed 303,000 capital construction projects and 300,000 modernization projects since 1979, increasing fixed assets by Rmb 829.5 billion. As part of the rectification programme, investment in fixed assets has been sharply reduced to Rmb 400 million in 1989, representing an 11 per cent fall in nominal terms; a 20 per cent decline in real terms when the increase in prices is taken into account. The differential ratio between supply and demand fell from 16.2 per cent in 1988 to about 8 per cent in 1989, as a result of decreasing demand for investment and slow increase in consumer demand.

A new source of investment since 1978 has been foreign capital which accounted for 6.25 per cent of total investment finance in 1990. Many overseas companies have helped to fund joint ventures, and China has borrowed extensively from international financial organizations. The way in which investment is channelled into enterprises have diversified as enterprises have begun to enjoy more autonomy; since then, local financial plans have become partially independent of Central Government control and financial resources and extra-budgetary funds used by local departments and enterprises have increased at a high rate. Thus, over the past ten years, State-owned enterprises boosted their budgeted investment by 46.8 per cent by raising funds themselves. As a proportion of total investment, this rose to 47.6 per cent in 1987 from 31.9 per cent in 1978.

The period 1983-1987 was distinguished by the very rapid growth of domestic loans which declined in 1988 and 1989. There was a relative decline in self-financing as an investment source in 1987. It rose significantly in 1988 and 1989, but fell marginaly in 1990 (see Table III.9). The switch to increased borrowing and to raising capital from external sources has been a significant contributor to the growth of inflationary pressure.

Industrial investment has generally averaged over 50 per cent of total investment in fixed assets during 1949-1988. The share climbed to almost 60 per cent during the period of reconstruction (1959-1962) and to 55 per cent during 1966-1976. The share of industry in total investment declined significantly during the first phase of the reform decade (1978-1982), when it averaged about 45 per cent. There has been a distinct upsurge with industrial investment growing at an annual average rate of 21 per cent during 1985-1988, which was almost double the rate of growth of total investment in fixed assets. During 1987 and 1988 industrial investment exceeded 51 per cent of the total.

Table III.9. Structure of investment finance, 1981-1990, selected years
(Percentage)

	1981	1983	1987	1988	1989	1990
State budget	28.07	23.75	13.06	9.1	8.26	8.71
Domestic loans	12.69	12.27	22.97	20.6	17.31	19.57
Foreign investment	3.78	4.65	4.82	5.7	6.63	6.25
Self financing	55.45	59.32	47.94	54.4	56.93	52.36
Other	..	..	11.23 a/	10.3 a/	10.88	13.10

Source: State Statistical Bureau of China, *China Statistical Yearbook*, various issues.

a/ An undefined proportion of this is in the form of foreign loan.

Fig.III.11. Sources of investment finance, 1981 and 1990 (Percentage)

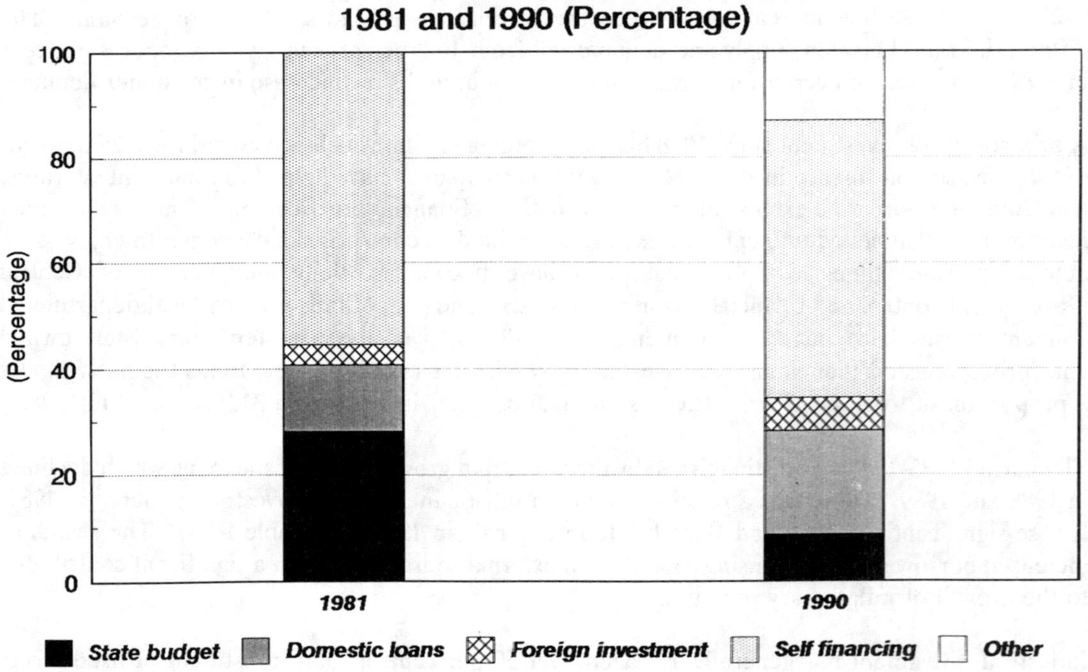

Industrial investment remains strongly concentrated in the heavy industries (see Table III.10) - chemical, metallurgical, petroleum refining, machine building - and the share of the unambiguously light industries has hardly increased at all. During the period 1981-1990, there has been a dramatic increase in the share of investment in power industry - this now accounts for 35 per cent of total industrial investment and this represents a major attempt at relieving what is perhaps the single most important bottle-neck constraining industrial performance. The petroleum segment of the energy industry accounted for around 11 per cent of industrial capital construction investment in 1990.

Table III.10. Distribution of industrial capital construction investment in major branches, 1981-1990
(Rmb million and percentage)

	1981-1985		1986		1987		1988		1989		1990	
	Rmb	per cent	Rmb	per cent	Rmb	per cent	Rmb	per cent	Rmb	per cent	Rmb	per cent
Petroleum	416.07	26.89	38.61	7.26	58.55	8.57	86.45	10.8	93.51	11.37	100.66	10.57
Machine building	156.17	10.09	41.75	7.85	50.17	7.34	56.82	7.1	63.21	7.69	73.38	7.70
Wood board	22.25	1.43	7.43	1.39	9.5	1.23	8.9	1.1	9.16	1.11	9.10	0.96
Building materials	78.51	5.07	31.88	6.00	34.16	5.03	32.10	4.0	28.19	3.43	29.1	3.05
Textiles	96.94	6.26	26.87	5.05	30.96	4.53	36.10	4.5	42.50	5.17	40.88	4.29
Food	64.01	4.13	23.31	4.38	30.71	4.49	29.46	3.7	28.47	3.46	26.75	2.81
Paper	10.76	0.69	4.32	0.81	5.36	0.78	6.73	0.84	5.95	0.72	6.22	0.65
Metallurgical	211.58	13.67	54.16	10.18	79.27	10.87	95.43	11.9	90.04	10.95	85.76	9.00
Chemicals	164.41	10.62	59.93	12.72	84.90	12.43	101.31	12.8	87.18	10.60	97.20	10.20
Total[a/]	**1546.97**		**531.64**		**682.79**		**796.09**		**822.48**		**952.60**	

Source: State Statistical Bureau of China, *China Statistical Yearbook* (various issues).

a/ Including power generation and coal mining.

Estimates of changes in fixed assets are of course subject to a wide margin of error. China has experienced substantial changes in both the level and structure of investment goods prices. Responding to the data requirements of an economy in which, until recently, resources have been allocated primarily through administrative rather than market mechanisms, Chinese statistical organizations have directed only limited effort towards the compilation of price indices, especially for intermediate and capital goods. Furthermore, available price indices may reflect official transfer prices, which in recent years have tended to diverge considerably from the market prices at which an increasing proportion of shares of intermediate and capital goods are now exchanged. When fixed asset values are deflated by these price indices the estimate for total fixed investment during the reform period is equivalent to roughly three-fourths of the total reported in Chinese sources. The growth rate of fixed investment was thus less spectacular than what appears to be the case at first glance.[13/] Furthermore reliable time series estimates of working capital employed by different categories of enterprises are also rare and hence estimates of total investment - and its growth - remain tentative. Despite these limitations it is evident that industrial investment has grown rapidly and has remained concentrated in the capital and intermediate goods branches.

Ownership pattern

There has been a marked proliferation of collective and town- and village-based enterprises (see Table III.11). In sectors such as building materials, metal products and furniture, small-scale enterprises account for about a quarter of the gross value of industrial output. They are also important in the textile and food manufacturing sectors. Many such enterprises - particularly in Shanghai, Tianjin and Guangdong provinces -have subcontracting relationships with State and foreign firms. Whereas subcontracting links with State firms have led to significant technological upgrading in the small-scale sectors, links with foreign firms have been of a short-term labour supply and processing nature, with little overall technological impact.

Collective industrial enterprises are much more subject to market discipline than large- and medium-sized State enterprises. A sample survey of 420 firms conducted in 1985 found that State enterprises sold 71 per cent of their output through official channels. This compares with a corresponding ratio of just 4 per cent for urban collective enterprises within the sample. State-owned enterprises received 84 per cent of their inputs from the State. As against this, urban collectives obtained only 6.5 per cent of their inputs from government allocations.[14/] During 1985-1988, the market procurement and the sales proportion of large-scale enterprises increased significantly but collectives are still well ahead of the former in terms of market orientation.

The growth of collectives has so far remained largely confined to the small-scale sector. This is also true of individually owned enterprises which currently account for over 4 per cent of the gross value of industrial output (see Table III.11). Since 1984, experimentation on a significant scale has occurred with share ownership schemes within large-scale enterprises, but progress remains limited.

Industrial investment was heavily financed by bank borrowing. Current liabilities as a proportion of net worth in State-owned companies increased from 11 per cent in 1978 to almost 80 per cent a decade later. Rising leverage ratios increase the financial risk of enterprises and discourage shareholding. The slow pace of share ownership growth has induced the government to restructure ownership through separating State and enterprise capital - the State enterprise is formally recognized as owning the capital which is financed through reinvestment of its retained funds. Enterprise autonomy seems to have increased due to these measures.

Township and village enterprises now number over 18 million, absorbing around 95.4 million of the rural labour force. Recently these industries earned $8.02 million foreign exchange. The value of output turned out by township- and village-based enterprises totalled Rmb 840.3 billion, an amount equal to China's gross product in 1979. Thus township- and village-based enterprises are playing an increasingly important role in changing the former mono-industrial structure of China. The fact that these enterprises have become an outlet for the surplus rural labour force was evidenced by a 62 per cent absorption of the increased labour force during 1978-1989.

The current phase of economic readjustments aims at rationalizing the product mix, improving the management and modernizing the enterprises. The introduction of the "Spark Plan" in the country's rural areas in 1986 came as a much wanted panacea (see Chapter IV). The "Spark Plan" emphasizes the establishment of rural industries based on agricultural and mineral resources. Within the framework of the "Spark Plan" scientific and technological projects are being carried out in the countryside in pursuit of introducing new technologies towards raising the quality of products and enhancing the efficiency of township- and rural-based industries.

Foreign investment

The opening up of the Chinese economy to foreign investment over the last decade ignited the aspirations of foreign firms to exploit the unlimited avenues of the huge Chinese market. Lured by China's abundant natural resource base and cheap labour over 8,000 contracts covering major forms of foreign direct investment - joint ventures, contractual joint ventures, wholly-owned ventures and corporative development of oil resources - were signed by end-1988. These forms of direct foreign investment committed foreign companies to make investments in cash, equipment, machinery and technology to the tune of $28.3 billion by the end of 1988.

Table III.11. Gross industrial output value by ownership, 1985-1990, selected years (Percentage)

Type of ownership	1985	1987	1988	1989	1990
State-owned enterprises	64.8	59.7	56.7	56.1	54.5
Collective, town- and village-based industries	32.1	34.6	36.1	35.7	35.7
Individual-owned enterprises[a/]	1.8	3.9	4.3	4.8	5.4
Other	1.2	2.0	2.7	3.4	4.3
Total	**100.0**	**100.0**	**100.0**	**100.0**	**100.0**

Source: State Statistical Bureau of China, *China Statistical Yearbook* (various issues).

a/ In urban and rural areas.

Fig.III.12. Ownership structure of gross industrial output value, 1990 (Percentage)

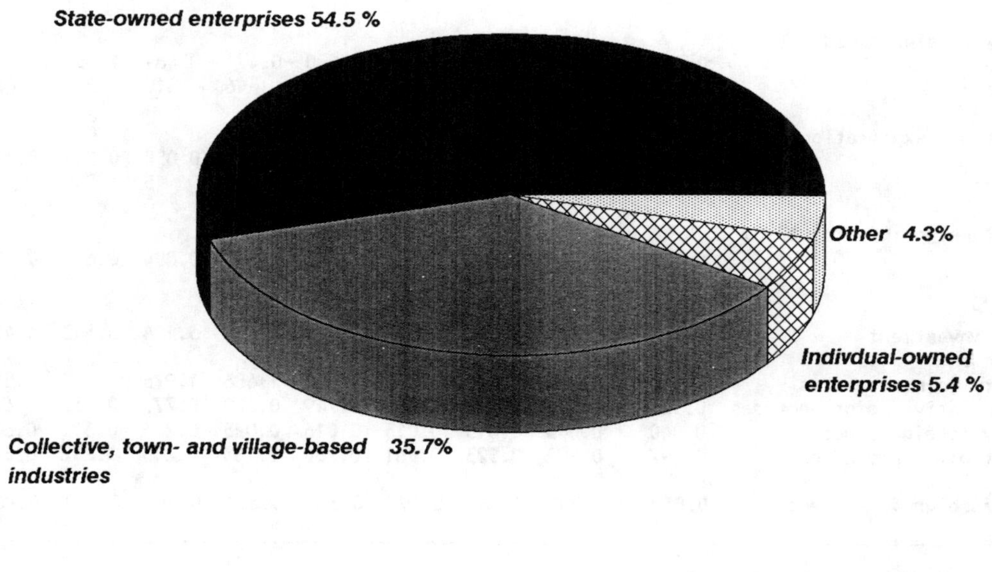

The turmoil of June 1989 brought a set-back in the flow of foreign investment. The value of negotiated investment reached $2.351 billion in the first half of 1990, representing a 22 per cent drop. China approved the establishment of 2,784 foreign-funded enterprises in the first half of 1990, down by 8 per cent from the same period in 1989. Of the 2,784 enterprises, 1,569 are in joint ventures, 570 co-operative enterprises and 645 solely foreign-owned enterprises.[15/] The contracted value of direct foreign investment in 1990 stood at $6.6 billion, while only $3.5 billion was actually utilized in the same year (see Table III.12).

The largest single category of direct investment hitherto has been co-operative joint ventures, accounting for almost half of the contracted value (see Table III.12). Guangdong continues to be by far the most popular site, and investors from Hong Kong and Macao are the most active (see Annex Table A-11). Guangdong's proximity to Hong Kong's capital and services and the ease of monitoring investment just across the border are major reasons. Besides Hong Kong and Macao investors, only firms from the United States and Japan have contracted to invest more than $1 billion. None of the other investors exceed $500 million. Investor interest in full ownership operations tends to increase and now it accounts for only 6 per cent of direct investment.

Table III.12. Type of foreign investment in China, contracted and utilized, 1979-1990
($ billion; units)

	1979-1982	1983	1984	1985	1986	1987	1988	1989	1990
Contracted									
Direct investment									
Value	4.608	1.731	2.650	5.931	2.834	3.709	5.297	5.600	6.596
Number of projects	922	470	1,856	3.073	1,498	2,233	5,945	5,779	7,273
of which:									
Equity joint ventures									
Value	0.127	0.188	1.067	2.030	1.375	1.950	3.134	2.659	2.704
Number	83	107	741	1,412	892	1,395	3,909	3,659	4,091
Co-operative joint ventures									
Value	2.727	0.503	1.484	3.496	1.358	1.283	1.624	1.083	1.254
Number	793	330	1,089	1,611	582	789	1,621	1,179	1,317
Fully foreign-owned									
Value	0.332	0.040	0.100	0.046	0.020	0.471	0.481	1.654	2.444
Number	33	15	26	46	18	46	410	931	1,860
Joint oil exploration									
Value	1.422	1.001	..	0.360	0.081	0.005	0.058	0.204	0.194
Number	13	18	..	4	6	3	5	10	5
Other foreign investment[a/]									
Value	0.927	0.185	0.224	0.402	0.496	0.610	0.894	0.694	0.390
Utilized									
Direct investment	1.168	0.635	1.258	1.658	1.875	2.314	3.194	3.393	3.487
Of which:									
Equity joint ventures	0.100	0.074	0.255	0.580	0.804	1.486	1.975	2.037	1.886
Co-operative joint ventures	0.531	0.227	0.465	0.585	0.749	0.620	0.779	0.752	0.674
Fully foreign-owned	0.040	0.043	0.015	0.013	0.016	0.025	0.226	0.371	0.683
Joint oil exploration	0.497	0.292	0.523	0.481	0.260	0.183	0.213	0.232	0.244
Other foreign investment[a/]	0.832	0.281	0.161	0.298	0.369	0.333	0.546	0.381	0.268

Source: MOFERT and *Business China*; State Statistical Bureau of China.

a/ Including international leasing, compensation deals, and processing and assembly.

Fig.III.13. Sector-wise distribution of direct foreign investment, 1990

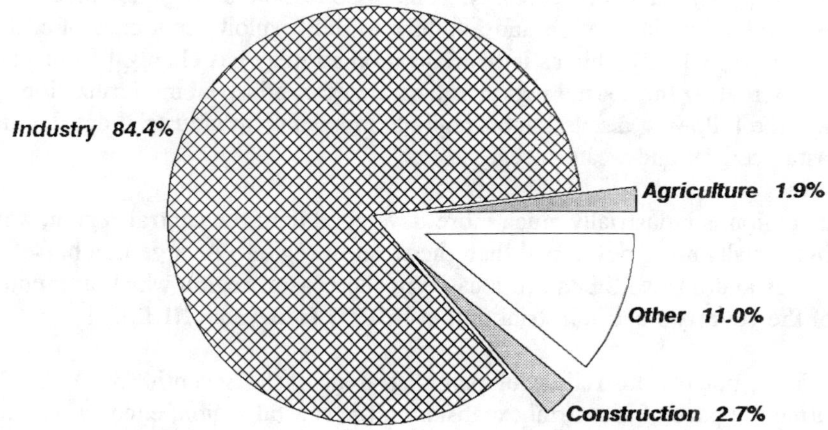

About 60 per cent of foreign investments were in manufacturing in 1987. Existing data do not report a breakdown of foreign direct investment by industrial branches. Wholly foreign-owned ventures generally invest three times more per production unit than equity or contractual joint ventures.[16/] In the mid 1980s over 70 per cent of equity joint venture projects involved foreign investment of less than $1 million and only 5 per cent had a foreign investment component of over $5 million. Most foreign-funded projects were located in the SEZs - Shenzhen alone accounted for over half of contracts involving foreign investment.[17/] Since 1986, foreign investment has also tended to grow rapidly in the coastal cities, but the share of the SEZs continues to remain dominant.

Given the small size of the typical foreign financial unit and the very high level of regional concentration, it is evident that the contribution of foreign enterprises to the modernization programme has necessarily been somewhat limited. This is a matter of considerable concern to the Chinese authorities whose primary interest has always been on impact on technological modernization and more recently on management upgrading.

It is evident that - by developing country standards - the incentive system is highly complex and provides ample opportunities for official scrutiny and regulation of the policies of foreign companies. Given the emphasis placed on the quality of investment a high level of government involvement is inevitable to provide the logistic and financial support which can enable transnational corporations (TNCs) to establish and sustain long-term technological and R & D collaboration with Chinese enterprise.

There are at present far too many national, provincial, municipal and local bodies cluttering up the system. Rules and procedures have been changed too frequently. Each SEZ has its own set of procedures and incentive packages - regulation and incentive procedures vary widely throughout the country. The Chinese authorities have themselves adopted a relatively "opportunistic" attitude *vis-à-vis* even major foreign investors. This is evident from the regulation that in principle each joint venture or wholly foreign-owned enterprise itself earns all the foreign exchange it uses in an accounting period. Similarly, restrictions on local borrowing are inflexible - they are rigidly fixed to the size of the foreign venture. Although the Eighth Five-Year Plan, beginning 1991, signals a further retreat from bold reforms already on hold since 1989, China endeavours to revamp the incentives system.

F. REGIONAL DIMENSION OF INDUSTRY

The regional dimension of industrial development in China is a pressing issue, given the uneven geographic distribution of the pace and pattern of industrialization. Until 1978, Chinese authorities placed great emphasis on the achievement of provincial self-reliance. This led to a relatively slow growth of factor mobility, a monopolization of regional markets by provincial enterprises, production duplication and a failure to fully exploit economies of scale. During the reform decade (1978-1988) China's industrialization strategy has changed from interior-oriented investments to treating the coastal region as a catalyst for industrial modernization. For analytical convenience, the following discussion on regional imbalance in industrial development is focused on the coastal, central and western regions.

The coastal region is industrially much more developed than the central region, while the central region is industrially more developed than the western region. A large number of China's major industrial cities and all five SEZs are located in the coastal region which accounted for over 50 per cent of the country's net industrial output in 1990 (see Table III.13).

The three-tier regional industrial development strategy of the Seventh Five-Year Plan was aimed at accelerating the pace of industrial expansion in the coastal region, encouraging energy and raw material development in the central region and making preparations for the development of the western region. The coastal region was expected to strengthen the technological transformation of traditional industries and develop knowledge- and technology-intensive as well as high value-added consumer products industries, while the central region would concentrate on producing energy and raw materials, selected machinery and electrical products and agricultural produce. The western region was hitherto destined for the development of agriculture, forestry, animal husbandry and selective development of energy, mineral resources as well as certain local processing industries. On the whole the emphasis was placed more on efficiency than on regional self-sufficiency. Although the rationale of this approach to regional development was based on the principle of comparative advantage, there has been both domestic and international investment bias in favour of the coastal region in the 1980s and the geographic dispersion of industrial production remained limited.

Four provinces - Shanghai, Jiangsu, Liaoning and Shandong - accounted for 36 per cent of the gross value of industrial production in 1987 - the figure was the same as at the beginning of the reform decade in 1979. The provinces which increased their share during 1978-1987 were Jiangsu (+3.3 per cent), Zhejiang (+2.7 per cent), Hubei (+1.3 per cent) and Guangdong (+1.1 per cent). As against this, the share of Shanghai and Liaoning - the traditional seats of heavy industry - went down by 2.4 per cent and 1.4 per cent respectively. The "gaining" provinces were often the seat of light industry and had succeeded in attracting foreign capital. This is particularly important in the case of Guangdong which has close economic links with Hong Kong and has been a principal beneficiary of the reform.

The most industrialized provinces - with a share of industry in net material product in excess of 50 per cent - are Beijing, Tianjin, Shanghai, Liaoning, Jiangsu, Shanxi, Heilongjiang and Hubei. The least industrially developed provinces are Guangxi, Xinjiang and Tibet. Nevertheless with the sole exception of Tibet, the share of industry in net material product exceeds 30 per cent throughout China. The industrial heartland is in the eastern and north-eastern regions and some southern provinces.

Amidst the apparent regional disparity in industrial development across the coastal, central and western regions, inter-provincial gaps in industrial performance have recently widened. During 1982-1990 six provinces registered a decline in their shares of net industrial output, twenty-one provinces experienced an increase and other provinces almost maintained their shares in the value of national net industrial output. The expanding provinces with significant increases in shares included Shandong, Fujian, and Guangdong and major declines in share were experienced by Shanghai and Liaoning. The expanding provinces are mostly middle-income provinces which are the seats of relatively light industry. The drastic decline in the share of Shanghai in net industrial output value is revealed by its fall from 14 per cent in 1982 to 8 percent in 1990. The least industrialized provinces have also experienced declines or marginal increases in relative shares during the 1982-1990 period (see Table III.13).

Table III.13. Net industrial output value by province, 1982-1990[a/], selected years
(Percentage of national total)

Province/Region	1982	1985	1988	1989	1990
Coastal region					
Liaoning	8.6	9.1	8.0	8.06	7.23
Beijing	4.7	4.2	3.9	3.88	3.73
Tianjin	3.8	3.4	2.9	2.69	2.70
Hebei	4.4	4.2	4.0	4.04	3.87
Shandong	5.9	7.4	6.7	6.83	7.04
Shanghai	13.9	11.1	8.6	8.07	7.98
Jiansu	5.6	5.6	5.7	5.67	6.16
Zhejiang	2.7	2.9	3.1	3.08	3.14
Fujian	1.5	1.6	1.9	2.04	1.99
Guangxi	1.6	1.6	1.9	1.96	2.03
Guangdong	4.1	4.4	4.8	4.78	5.08
Hainan					0.24
Central region					
Jilin	2.5	2.8	3.3	2.97	3.00
Heilongjiang	6.8	5.8	6.8	6.55	7.15
Shanxi	2.9	2.5	2.5	2.58	2.54
Nei Monggo (Inner Mongolia)	1.4	1.4	1.5	1.58	1.60
Anhui	2.6	2.7	2.8	2.72	2.72
Jiangxi	1.6	1.9	2.1	1.99	1.90
Henan	4.0	4.3	4.5	4.42	4.54
Hubei	4.4	4.9	5.0	4.89	4.57
Hunan	3.3	3.3	3.7	3.75	3.61
Western region					
Shaanxi	2.2	2.2	2.3	2.45	2.45
Gansu	1.9	1.9	1.8	1.78	1.88
Qinghai	0.3	0.3	0.4	0.46	0.40
Ningxia	0.3	0.3	0.3	0.41	0.39
Xinjiang	0.9	1.1	1.3	1.29	1.33
Sichuan	5.2	5.7	5.9	6.40	5.86
Guizhou	1.1	1.4	1.6	1.64	1.69
Yunnan	1.8	1.9	2.4	2.75	3.13
Xizang (Tibet)	..	..	..	0.03	0.03
Total	**100.0**	**100.0**	**100.0**	**100.0**	**100.0**

Source: State Statistical Bureau of China.

a/ Data presented in this Table are for State-owned independent accounting industrial enterprises.

The expanding provinces are clustered towards the east of the country and are the domain of newly emerging collective enterprises. The east coast provinces have better infrastructural facilities and have also witnessed a rapid proliferation of industries. However, the most industrialized provinces - Jiangsu, Shanghai, Liaoning and Beijing - predominate in both light and heavy industrial branches. As against this, the relatively underdeveloped areas of Sichuan and Shandong primarily encompass light industries.

The coastal region's share of investment in fixed assets has increased gradually since the late 1970s. In the early 1980s the coastal region absorbed about half of gross fixed investment, rising from around 40 per cent during 1953-1980.[18/] This indicated a significant shift in investment strategy in favour of the coastal region. In 1990 over 50 per cent of investment in fixed assets in industry was directed to the coastal region, while less than 20 per cent was destined for the western region which encompasses the industrially least developed provinces (see Annex Table A-12).

Most of the foreign investment also went into the coastal region during the reform period, with Guangdong alone hosting over 65 per cent of foreign investment enterprises in China (see Table III.14). Three of the five Special Economic Zones are located in Guangdong, which is being called Asia's "fifth dragon".[19/] During the period 1979-1990 the coastal region as a whole hosted over 90 per cent of foreign investments, while around 5 per cent of foreign investments was attracted by the central region and less than 3 per cent of foreign investment was directed to the western region. Of the total number of 274 foreign investments in the western region, 100 investments were hosted by Shaanxi during 1979-1987. In 1990 Sichuan emerged as the most important host of foreign investment. Of the total number of 208 foreign investments in the western region in 1990, Sichuan hosted 134 investments.

The government's intention is to transform the coastal region into a processing centre for inland raw materials. Endowed with relatively higher levels of technology and managerial skill, the coastal region is expected to absorb the imported advanced technology and to develop indigenous capabilities for technical and economic innovations. Having achieved this, the coastal region is expected to diffuse the technology throughout the Chinese economy.

The fact that the coastal region attracted a greater investment flow than the central and western regions is evidenced by the number of enterprises established and the value of gross industrial output generated by the different regions (see Annex Table A-13). However, industrial concentration is not determined by the number of enterprises alone. For example, Sichuan, located in the western region, encompassed as many as 781,140 industrial enterprises in 1990, but contributed around 5 per cent to gross output value. The largest number of industrial enterprises are located in Henan, while the highest contribution to China's gross industrial output stems from Jiangsu which had 630,020 enterprises, compared with 810,800 in Henan in 1988.

The almost imperceptible change in the provincial distribution of industrial production reflects the ability of provincial and municipal administrations to protect and further their industry. The policy-making powers of provincial and municipal administrations have been significantly augmented due to the centre's commitment to decentralization of decision-making structures. Provincial and municipal administrations have often proved themselves adept at pre-empting the autonomy conceded to enterprises and at formally reconstituting provincial bureaucratic structures into State level enterprises. This reconstitution has been achieved without any substantive change in operational management procedures. Provinces have also often successfully resisted the erosion of barriers which restrict commodity and factor mobility. Provincial administrations are reluctant to reduce excess capacity, specially in the heavy industries. There is a particularly urgent need to rationalize production structures located in remote areas.

The spatial reorganization of industry to better reflect regional comparative advantage remains problematic politically. This is particularly exacerbated by the low volume of inter-regional trade. While the prospering coastal and eastern provinces have developed extensive trade links with the Pacific Basin, inter-provincial trade levels and marketing ties remain undeveloped.

Market forces have a very limited role in the provinces of the interior which remain dependent on the heavy defence-related industries established and controlled by the State. The access of interior and western provinces such as Xinjiang and Sichuan to foreign exchange is very limited. Although formally a dual price system operates in both coastal and interior provinces, in the latter, market prices are usually strictly regulated by the State. Collective industrial enterprises remain very weak within the interior provinces. Despite the slower growth of interior provinces, wage structures are broadly similar, accentuating the competitive advantages of the coastal provinces. The coastal provinces have better access to capital, with greater autonomy to retain surplus, and the growth of regional autonomy has meant that it is the coastal provinces which are in a better position to raise both domestic and foreign loans.

Table III.14. Foreign investment by province and region, 1979-1990

Province/ Region	Number[a] of investments				Value of investment ($ million) Contractual				Actual			
	1979-1987	1988	1989	1990	1979-1987	1988	1989	1990	1979-1987	1988	1989	1990
Coastal region												
Liaoning	212	272	271	371	779.2	192.3	301.1	493.9	145.0	115.3	118.6	243.7
Beijing	261	148	185	241	1,651.2	143.2	79.4	117.7	640.0	502.8	318.5	277.0
Tianjin	230	94	97	129	242.8	109.6	84.5	131.7	150.4	31.9	28.0	34.9
Hebei	103	92	73	110	98.4	186.9	63.1	85.9	19.5	16.7	26.9	39.4
Shandong	126	203	240	366	336.0	260.2	178.6	232.8	..	43.1	131.3	150.8
Jiangsu	194	247	277	393	219.9	302.0	192.9	285.6	88.9	103.0	93.6	124.2
Shanghai	291	219	199	201	1,842.6	333.3	359.8	374.6	366.4	233.2	422.1	174.0
Zhejiang	153	152	184	294	158.7	113.3	83.6	133.1	62.6	29.6	51.8	48.4
Fujian	1,023	813	872	1,043	821.9	462.6	902.6	1,161.8	300.5	130.2	328.8	290.0
Guangdong	6,970	2,706	2,438	3,042	9,188.3	2,242.0	2,438.1	2,689.6	3,112.4	957.9	1,156.4	1,460.0
Guangxi	287	122	77	114	384.1	112.7	25.3	125.2	120.0	20.7	45.9	28.7
Central region												
Heilongjiang	97	49	60	76	103.0	58.1	47.2	27.1	13.2	40.1	22.4	24.5
Jilin	45	25	44	52	95.1	9.1	22.6	20.8	..	6.2	3.4	17.6
Inner Mongolia	15	14	6	14	48.8	6.2	8.2	19.5	12.6	3.4	0.2	10.6
Shanxi	19	20	19	25	5.7	15.5	12.5	12.0	2.8	6.5	8.8	3.4
Henan	78	38	36	50	224.8	22.9	16.8	21.1	16.4	64.2	42.7	10.5
Anhui	54	35	34	53	1.9	10.9	6.1	19.2	..	11.5	4.8	9.6
Hubei	78	53	70	99	55.9	78.0	47.8	39.5	36.1	22.3	23.0	29.0
Jiangxi	66	35	24	54	74.1	17.6	5.1	28.6	14.5	5.2	5.9	6.2
Hunan	99	22	37	45	67.5	14.3	18.8	26.8	9.7	7.7	6.4	11.2
Western region												
Xinjiang	17	8	9	6	62.0	22.4	2.8	58.3	34.0	5.0	0.9	5.4
Gansu	19	6.0	5	8	23.6	13.3	4.5	2.9	4.1	2.0	..	0.9
Ningxia	2	3	7	4	2.0	1.6	7.0	1.0	..	0.3	1.1	0.3
Shaanxi	110	13	23	24	1,029.2	14.5	31.1	11.3	12.2	111.7	97.2	41.9
Qinghai	4.0	..	..	..	4.2	..	..	...	0.5	2.7	..	...
Sichuan	73	34	33	134	140.7	25.4	21.2	100.6	42.9	23.6	8.0	16.0
Guizhou	24	13	15	21	12.8	15.1	7.3	16.3	3.7	4.4	7.5	4.7
Yunnan	23	8	10	11	20.6	11.0	3.0	2.5	11.1	3.1	7.4	2.6
Tibet	2	1	..	..	1.2	0.03	..	...	1.2	0.03	..	...

Source: *Shanghai Industrial and Economic Journal*, 8 July 1988, State Statistical Bureau of China.

a/ Including joint ventures, co-operative enterprises, and wholly foreign-owned subsidiaries.

With a view to mitigating to some degree the regional imbalances in industrial development, the central government encourages domestic joint ventures, compensation trade, investment in the interior by companies from the coastal region and technical exchanges between provinces. The current imbalance in the industrial structure is attributed in part to the failure of central and western regions in terms of supply response to the rapidly growing coastal region. Inadequate compensation to enterprises in the interior is one of the reasons for the insufficient supplies of raw materials to the coastal region. Given the political complexity and the absence of diffusion effects of the coastal "growth poles", regional imbalance in industrial development continues to characterize the Chinese pattern of development. Inter-provincial co-operation through investment redeployment as well as the transfer of technology from the coastal to the interior is deemed pivotal to speed up the transmission of industrial development and the inter-provincial division of labour and specialization.

G. SELECTED INDUSTRIAL LOCATIONS

Guangdong

Guangdong Province, located in the southern part of China, near Hong Kong and Macao, with a population estimated at 64 million in 1989, has three of the five Special Economic Zones - Shenzhen, Zhuhai and Shantou. Its capital Guangzhou (Canton) enjoys the status of an especially privileged open city. Guangdong has grown rapidly in the recent past and has been the centre of the foreign investment influx.

The province's agricultural resource base encompasses rice, sugarcane, fruits, tea, rubber, silk-wear, cocoon and livestock products. Endowed with over 4,000 kilometres of coastal line, fishing is well developed. The province is rich in mineral resources, with 85 commercially viable proven mineral resources, especially non-ferrous metals.

Guangdong is the seat of light industry such as food manufacturing, textiles, wearing apparel, paper, printing, arts and crafts. The electrical and electronics branches are also dominated by consumer durables production. In 1987 light industry accounted for 62 per cent of the gross industrial output value unambiguously. The predominance of light industry is also reflected in the relatively greater importance of collectives and individually owned enterprises in Guangdong. Together they produced 42 per cent of gross industrial output, representing one of the highest provincial shares in China. In Guangdong, TNCs have established fully owned subsidiaries and have to a limited extent entered into contracts with collectives. However, the major TNC partners remain State-owned enterprises.

Capital intensity is high and labour productivity has grown rapidly within Guangdong's industrial sector in both State-owned enterprises and others. Despite this, profitability levels are modest. In 1987, the surplus realized for every Rmb 100 of fixed asset value was only Rmb 23.5 for Guangdong. This was significantly higher than the national average (Rmb 19.72) but well below the investment return rates in Shanghai, Zhejiang, Beijing, Tianjin and Jianghan. This partly reflects the heightened competition facing Guangdong's industrial sector. The decline in the surplus rate in recent years also partly indicates an exhaustion of the most profitable investment opportunities in Guangdong. Investment and production costs have also escalated rapidly in the past three years.

Nevertheless investment has grown rapidly in recent years due largely to Guangdong's close ties with neighbouring Hong Kong. Around 90 per cent of investment in the 60,000 companies, which are either wholly or partly owned by foreign concerns, originated from Hong Kong which absorbs about 60 per cent of Guangdong's exports.[20/] Around 80 per cent of commodity prices are being determined by market forces.

Guangdong operates the most liberal foreign trade and investment regime in China. Restrictions on the use of foreign exchange earnings by companies are few. Guangdong is the centre of

China's trading network - the leading export fair is held every year at Guangzhou which generates about a quarter of China's export orders.

The Guangdong International Trust and Investment Corporation (GITIC) is the main vehicle for attracting foreign investment. This is one of the largest subsidiaries of the China Investment and Trust Corporation and can authorize investment up to Rmb 30 million.

Investment priorities for Guangdong Province for 1990 were as follows:

- Technological refurbishing of existing enterprises and development of traditional and new industries:

(a) food, beverage and packaging industries;
(b) dyeing, textile and garment industries;
(c) computer, electronic and domestic electrical appliances;
(d) sugar and paper industries;
(e) refrigerating and air conditioning industries; and
(f) others including pharmaceuticals, rubber-ware, paints, lithophone, titanium white, leather, leather goods and electroplating.

- Development of new materials and new products:

(a) aluminium plates, colour coated and enamel coated steel plates;
(b) chemical fibres, dacron silk reeling, synthetic fibres, worsted spindles and woollen goods;
(c) fertilizers and soda ash; and
(d) electrical appliances, instruments, wires, cables, chemicals, machinery and accessories, wallpaper, rugs, carpets and other items for interior decoration.

- Development of new high technology industries such as computer software and accessories, integrated circuits, multilayer circuit boards, sensitive devices, sensor devices, glues, organic pigments, optical fibre communications, engineering pottery, amorphous silicon, solar cells, biological and enzymes engineering.

An important focus of investment activity is Guangdong's capital city Guangzhou (Canton). The "Open Door" policy has given Guangzhou the opportunity to take advantage of its strategic situation in foreign trading, as it stands at the apex of a triangle whose base extends from Hainan Island in the west to the borders of Fujian Province in the east, and it draws on a large area for resources.

The city's industrial strength is based on the considerable natural resources of the Pearl River and its hinterland, taking advantage of the financial and commercial experience of Hong Kong and Macao. Fertile plains and an abundant labour force contribute to the economic base on which Guangzhou can draw.

Guangzhou has established direct trading relationships with more than 140 countries and regions. In addition to an array of trading corporations, the city has its own packaging and advertising corporations. The trading corporations are also involved in compensation trade and joint ventures. Under the "Open Door" policy of China, these corporations have been given the flexibility of establishing business relationships around the world.

Shanghai

The Shanghai municipal area, with a population of over 12 million, is the second largest agglomeration in China. Seven and one-half million people are concentrated in the main urban region. The population density is about 3,200 per square kilometre, the highest in China. Shanghai is the country's main port with well over 50 per cent of internal and external trade passing through the city.

Shanghai is well endowed with natural resources. It has a skilled labour force and an innovative managerial cadre imbued with industrial mentality. The physical infrastructure is relatively well developed and the supply of industrial raw materials is ensured by the high volume of trade and by extensively mechanized agricultural production in the surrounding region. Water resources - from Chengjiang river and Lake Jaihui, as well as from underground sources - are plentiful. However, mineral resources are insignificant with the exception of the coal reserves currently being jointly developed with other provinces.

The major branches are machine building, textiles, metallurgy (especially non-ferrous and metal products), electrical and electronics engineering, transport equipment (especially shipbuilding) and food manufacturing - together these branches accounted for over 50 per cent of gross industrial output value in 1987. Despite the relatively rapid growth of collectives, State enterprises continue to remain dominant within Shanghai's industrial structure - they accounted for almost 80 per cent of gross industrial output value in 1987. The vast majority of joint venture and licensing agreements with TNCs are established by State enterprises. The State sector registered a decline in its output and a sluggish growth rate of 1 per cent in labour productivity in 1989. Booming rural industries now account for most of the increase in output and employment. The collectives and private enterprises currently set the pace of industrial expansion.

Light industry accounted for 56 per cent of gross industrial output value in 1985. By 1987 this share had fallen to around 50 per cent reflecting the very rapid recovery of heavy industrial production since 1984. Given the predominance of the State sector, industry in Shanghai naturally exhibits a higher level of capital intensity. Moreover, rates of return on industrial investment are among the highest in China - all independent accounting units earn on average twice the rate of surplus per unit of investment that is realized nationally. State industrial enterprises obtained a surplus rate that was 94 per cent higher than the national average. While the initial cost of establishing and maintaining investment is certainly higher in Shanghai than in the neighbouring east coast provinces, the average rates of return remain significantly higher.

Shanghai had the highest provincial share in modernization investment - 7.6 per cent - during 1985-1987, followed by Guangdong, Sichuan, Jiangsu, Beijing and Shandong. Industrial investment in Shanghai remains concentrated in a relatively small number of branches - power generation equipment, chemicals and ferrous metal products accounted for about 75 per cent of capital construction expenditure in 1987. Modernization investment was concentrated in textiles, chemicals, ferrous metal products, machine building, transport equipment and electronics, which accounted for 60 per cent of such investment in 1987.

New industrial products introduced in recent years include metallurgical materials, high polymer compounds, precision instruments and metres, computers, petrochemicals and precision machine tools. Shanghai accounts for 40 per cent of the national output of television sets, wrist watches and radios and over 20 per cent of the national output of sewing machines, chemical fibres, ethylene, woollen fabric and tape recorders. Major export products from Shanghai include rolled steel, machine tools, sewing machines, woollen fabrics, cotton yarn, silks, toys, bicycles, paper and cotton cloth - these are mainly products of light industry which is much more export-oriented than the heavy industrial units.

Branches tend to be concentrated in industrial estates and satellite towns - petrochemicals in Gaochea and Jinshian, chemicals in Tiapou and Miabriang, electronics in Chaobeijian and Anting, automobiles in Anting, light industry in Sonjiang and Changgiao, metal products in Wassong and Songjiang.

During 1985-1988 Shanghai's industrial growth - measured in both output and export terms - has been significantly lower than that of other east coast provinces such as Jiangsu and Guangdong. This was partly explained by the growing strain on the physical infrastructure and the fact that modernization expenditure has been insufficient to adequately replace rapidly ageing capital stock. The predominance of State enterprises and the emphasis that the Shanghai local administration has placed on maintaining a centralized control structure has also been a factor inhibiting the reckless growth in industrial investment that has characterized Guangdong in recent years. Moreover, investible resources have also been depleted by Shanghai's extraordinarily high federal revenue delivery obligations.

During the Seventh Plan period, the Shanghai municipal authorities have placed emphasis on a co-ordinated development of industry, transport, construction, commerce and social service. Rapid introduction of new and upgraded technology has been encouraged and seven new industries are to be developed on a priority basis. Microelectronics, lasers, fibre optics, communications, biotechnology, marine engineering and robotics, shipbuilding, aircraft, telecommunications and petrochemicals were earmarked for rapid technological modernization during 1986-1990. Industries are being encouraged to develop an export orientation. The domain of Shanghai's foreign trade authority has been extended to include surrounding satellite towns. In order to facilitate the spread of industry from the centre to the suburb infrastructure - especially transport and communications - investment has grown rapidly. During 1985-1987, an annual average of Rmb 1 billion was spent on infrastructural investment in Shanghai. A new north-south expressway has recently been constructed to facilitate road transport. A second tunnel across the Husanengpu river is nearing completion. The road to the airport has been upgraded. A new passenger station in East Shanghai is being constructed. This will link all north-south and east-west trains in China. This is a very important development and will lead to doubling of train services to Shanghai (expected to be about 150 daily by the early 1990s).

Water transport - which is Shanghai's most important means for freight carrying - is also being developed. The port, which ranks among the ten largest in the world, is being expanded to cope with an estimated annual volume of 180 million tons by 2000. New berths are being developed at first in neighbouring cities.

Shanghai continues to face an acute energy shortage. Its main sources of energy are hydroelectric and coal fired thermal plants. Coal-fired power plants with aggregate generating capacity of over 1,200,000 kW have been constructed during 1985-1988.

Since 1985 Shanghai has been designated as one of the 14 major coastal "open" cities. Tax reductions are applicable to approved high technology projects as well as profits in excess of $30 million. Large manufacturing projects have been established by Pilkington, V.W. Schneider and MacDonnell Douglas. The Shanghai Municipal Foreign Economic and Trade Commission is the sole approved authority for all joint venture projects (except those needing special approval).

During the reform decade Shanghai has also emerged as a major technology importer. Technology imported relates both to the establishment of new industries - such as microelectronics, lasers, optics and biotechnology - and the modernization of existing ones. Technology imports are also a means for the construction of new development zones such as those located in the Minbing District and Hanggiao District. The Minbing District has been developed to attract foreign investors. It is hoped to attract small- and medium-sized firms using modern technology producing electronics, instruments, toys, meters, garments, food manufactures, domestic appliances, etc. Hanggiao has been developed as a pleasant residential district for foreigners.

Shanghai has the reputation of being one of the most liberal and generous foreign investment regions. Tax treatment is almost as favourable as that offered by the SEZs. Tax on profits can be reduced to 10 per cent in specified cases. Such projects can import construction material, raw materials, component parts and packing material free of customs duties and commercial taxes. Exports of many products are tax exempt and a proportion of the output of joint ventures can be sold on the domestic market. Joint ventures management boards have extensive autonomy in determining employment policies.

Special Economic Zones and Open Cities

China repeatedly reiterates that the doors remain open to foreign investors, and efforts are under way to further improve the investment environment especially in the SEZs and open cities.

The most noteworthy aspect of the five Special Economic Zones - Shenzhen, Zhuhai and Shantou in Guangdong Province, Xiamen in Fujian Province and the island Province of Hainan - has been an accelerated export drive. In 1989 export value of SEZs totalled $3.85 billion, representing almost one-tenth of China's total exports. The five SEZs also achieved substantial gains in absorbing foreign investment. By the end of 1988, the five SEZs have approved foreign-funded projects to the tune of $9.4 billion. Actual implementation of $4.1 billion-worth of projects represented one-fourth of the country's total. Data for 1990 show that the SEZs approved 2,124 foreign-funded projects worth of $1.9 billion, of which $1.1 billion was utilized (see Table III.15). Of the 14 coastal open cities, Guangzhou's approval of 2,121 foreign-funded projects in 1989; surpassed that of any Special Economic Zone and of all other open cities. Guangzhou utilized over 50 per cent of the contractual value of foreign-funded projects in 1989. In 1990, the coastal open city Fuzhon approved the largest number of foreign-funded projects.

Table III.15. Utilization of foreign funds by SEZs and open coastal cities, 1990

SEZ/City	Number of foreign-funded projects	Contracted value (million $)	Actually utilized value (million $)
Special Economic Zones (SEZs)			
Shenzhen[a/]	763	689.61	518.57
Zhuhai[a/]	407	293.39	108.28
Shantou[a/]	340	258.42	158.53
Xiamen[a/]	362	565.63	173.07
Hainan[a/]	252	128.82	100.65
Coastal open cities			
Dalian	213	710.43	394.17
Qinhuangdao	15	6.89	19.39
Tianjin	131	185.59	332.96
Qingdao	65	75.11	75.19
Yantai	87	86.11	52.53
Weihai	41	20.73	8.20
Shanghai	222	571.60	779.70
Nantong	34	50.11	38.35
Lianyungang	12	10.61	12.15
Ningbo	93	76.46	77.03
Wenzhou	42	10.95	2.50
Fuzhou	233	272.70	102.20
Guangzhou	392	522.50	267.37
Zhanjiang	59	50.55	25.08
Beihai	20	36.74	10.52

Source: State Statistical Bureau of China.

a/ The data here is not only for SEZ area.

Although the SEZs and open cities continue to constitute the "windows" of China's "Open Door" policy, a close look at these especially privileged regions reveals that the "window function" has failed to create linkages with the larger national economy.[21/] An assessment of SEZs and open cities should be viewed within the perspective of the envisaged goals.

One of the objectives of SEZs and open cities is to promote foreign investments. So far, around 80 per cent of foreign investments originated from Hong Kong and Macao. These zones have not been successful in attracting potential investors from other parts of the world. The firms in SEZs seem to be an extended workbench of their parent companies in Hong Kong and Macao and to some extent Taiwan Province because of the relatively favourable real estate prices and low wages in China.

A major objective of creating SEZs and coastal open cities was further to attract foreign technology, with a view to achieving spill-over effects for the modernization of the Chinese economy. A large number of factories are labour-intensive, using conventional technologies. These firms process mostly imported intermediate products or assemble imported components. As relatively low wages still remain a major attraction, foreign firms do not generally opt for modern, automated, high capital-intensive production devices. In the structure of industrial output in Shenzhen, for example, electronics accounted for 43 per cent by the late 1980s. Although direct and indirect spill-over effects on other regions do not seem to be evident, the SEZs and open cities have been at the forefront of economic exchange and co-operation between China and the international industrial community. They have undoubtedly played a much-needed role as industrial gateways of China. These coastal vanguards could play a significant role in mitigating the recentralization trends and thereby overcome uncertainties surrounding the industrial investment environment in China.

Fig.III.14. Contracted and utilized value of foreign-funded projects in five SEZs, 1990 ($ million)

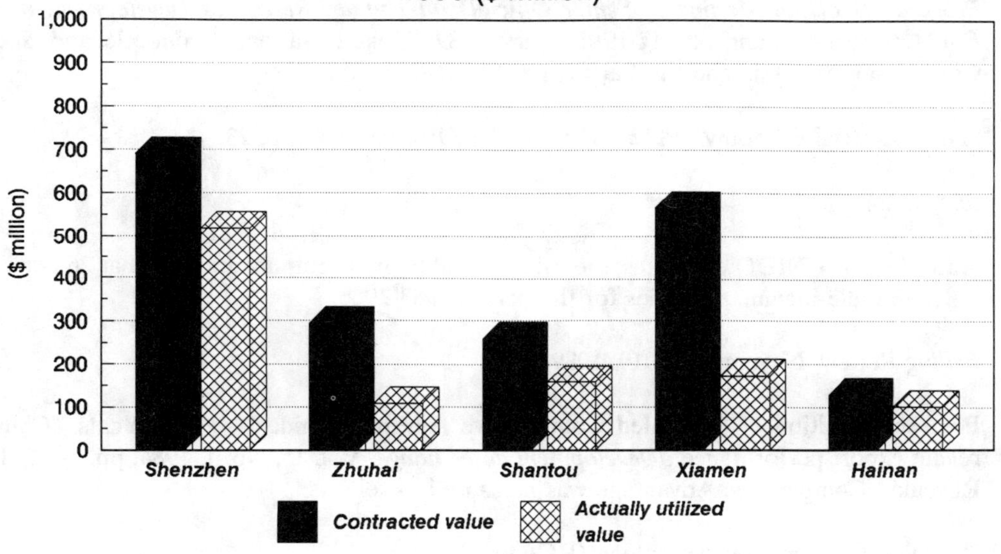

Fig.III.15. Contracted and utilized value of foreign-funded projects in selected coastal open cities, 1990 ($million)

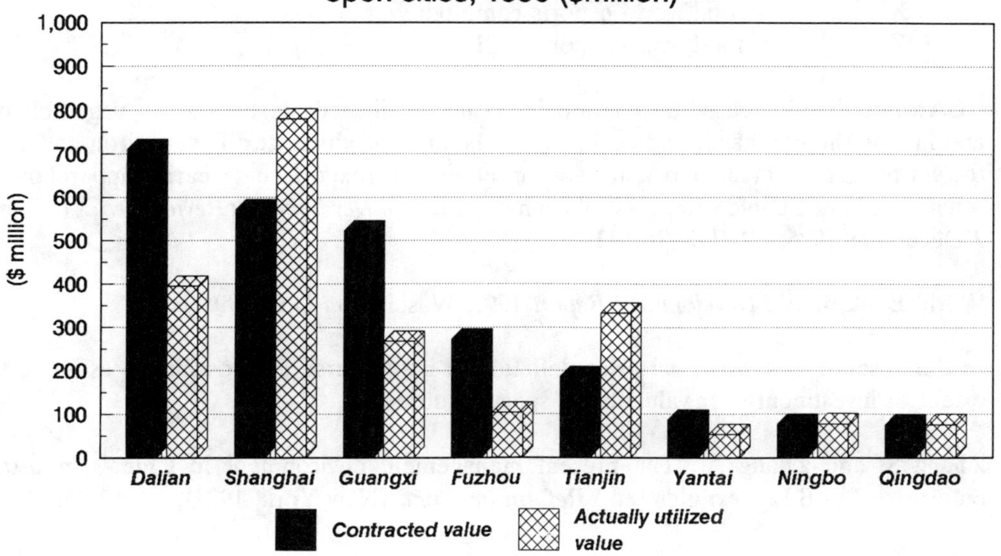

NOTES TO CHAPTER III

1/ See World Bank, *World Development Report 1990* (Washington D.C.), pp. 188-189.

2/ "Heavy" and "light" industry categories are not entirely unambiguous in Chinese industrial statistics since chemicals and wood processing are common to both categories.

3/ Chinese statistics do not furnish disaggregate information on the overall structure of manufacturing employment.

4/ For an in-depth analysis of sources of productivity growth, see Gary H. Jefferson, "Potential sources of productivity growth within Chinese industry" *World Development* (United Kingdom, 1989), Vol. 17, No. 1, pp. 45-47.

5/ *Beijing Review*, July 2-8, 1990, p. 21.

6/ This section draws largely on the information provided in Zhang Dengyi, "China" in *Proceedings of the Meeting of Senior Officials on Human Resources Development in the ESCAP Region*, Bangkok, 17-19 January 1989, United Nations Economic and Social Commission for Asia and the Pacific, pp. 147-168.

7/ "The industrial economy", *Beijing Review* (3-8 October 1988), p. 23.

8/ *Ibid*, p. 24.

9/ According to UNIDO estimates, the value added to gross output ratio stood at 26 per cent for 53 middle-income countries for the period 1983-1986.

10/ *Beijing Review*, November 13-19, 1989.

11/ For the definition of "Revealed Comparative Advantage Index" see, A. Yeats, "China's recent export performance", *Development and Change*, Vol. 15, No. 1, 1984, pp. 1-22. The Revealed Comparative Advantage was measured as follows:

Revealed Comparative Advantage (RCA)

RCAk = (Xik / Xi) / (Xk / X)
Xik = country i's exports of commodity k
Xi = country i's total exports
Xk = total world exports commodity k
X = total world exports (SITC 0-9)

RCA values had increased for commodities such as oil-seeds, textile fabrics, vegetable oils and fats on the one hand, and declined for chemical products, textile yarn, iron and steel, the entire range of machinery and transport equipment, exports and wearing apparel on the other. See "The People's Republic of China", *Industrial Development Review Series* (Vienna, 2 December 1985, UNIDO/IS.582).

12/ World Bank, *World Development Report 1991*, Washington D.C. 1991.

13/ In the absence of subsectoral price deflation, it is of course not possible to estimate the extent of investment over valuation in specific subsectors.

14/ Zhang S. and Zhang A. "The present management environment in China's industrial enterprises" in B.L. Reynolds (ed.) *Reform in China* (New York, 1987), pp. 52-56.

15/ *Beijing Review*, (August 13-19, 1990), p. 28.

16/ The distinction between equity and contractual joint ventures has always been a relatively arbitrary one. In principle contractual ventures are more flexible (it is not subject to the China Law on Joint Ventures), nevertheless most contractual joint ventures do involve equity ownership by the foreign partner.

17/ Data in this paragraph are sourced from World Bank, *China: External Trade and Capital* (Washington 1988), pp. 256-258.

18/ For an in-depth analytical interpretation of changing regional development strategy see Dali Yang, "Patterns of China's regional development strategy", *The China Quarterly* (Oxford, June 1990), pp. 230-257.

19/ See Louise do Rosario, "Asia's fifth dragon", *Far Eastern Economic Review* (8 December 1988), Vol. 142, No. 49, p. 62.

20/ *South* (April 1989), p. 37.

21/ For a critical appraisal of the SEZs, see Petra Pissula and Dieter Loesch, "Special Economic Zones in the People's Republic of China" *Intereconomics* (September/October 1990), pp. 257-264.

IV

INDUSTRY BRANCH PROFILES: RETROSPECTS AND PROSPECTS

A. FOOD PROCESSING: ENHANCING COMPETITIVE ADVANTAGE

The resource base

The agricultural resource base supporting the food processing industry is substantial in China. The main food grains are rice, wheat, sweet potatoes, maize, sorghum, millet and soya beans. Grain production experienced the fastest rate of expansion during 1978-1984, despite a drastic decline in 1980 in the wake of natural disaster. Farmers' enthusiasm was sparked by the introduction of a contract responsibility system as well as remuneration linked to output. Since 1979 the government has significantly increased the purchasing price for grain and paid 50 per cent more for any above-quota purchases. This has served as a great incentive. Given the low cost of inputs, comparative earnings from grain production rose markedly during 1978-1984. Despite a reduction in sown area productivity gains significantly boosted the output of food grains. Following the peak level of output in 1984, the period 1985-1988 was characterized by erratic growth, with no clear trend in grain production. A significant rise in the costs for the means of agricultural production resulted in a sharp drop in the comparative earnings from grain production. An overall reduction in agricultural investment and neglect of agriculture in finance and credit allocations also led to stagnation in grain production.

China endeavours to raise its grain output from the annual volume of 435 million tons in 1990 to 500 million tons by the year 2000. Farmers are encouraged to make greater use of new technology in order to enhance productivity, both through increased reliance on multiple crop planting and through the use of inter-cropping and inter-planting. China's well-established research and application system for nuclear agronomy is expected to achieve improvements in crop variety, pest prevention and control, and food preservation.[1/]

China's rice production stood at 189.33 million tons in 1990, compared to 178.26 million tons achieved in 1984 (see Table IV.1). As the sown area for rice production is likely to remain unchanged in the 1990s, production increase has to come from higher yields. Plans are under way to improve rice yields in low yield areas where considerable potential seems to exist. A major constraint to expanding the area for hybrids is the relatively long period for maturation. Future advances in rice yield hinge on the extensive use of hybrids with a shorter maturation period. Improvements on the rice research front seem to augur well for significant productivity gains in rice production.

The production of wheat rebounded from 85.90 million tons in 1987 to 98.23 tons in 1990. Despite significant increases in the production of wheat in 1989 and 1990, China continues to depend heavily on imports in order to meet the growing demand. Annual growth in wheat yields is higher than the world average, but lower than some agriculturally advanced countries. Amidst erratic growth trends in the production of corn there has largely been production stagnation until 1989 (see Table IV.1). A significant increase in the production of soya beans stemmed largely from increased yields over the years. However, Chinese yields remain far below average yields in developing countries.

Of all oil-bearing crops peanuts and rapeseed rank top in terms of sown area and output. In 1990, China produced 6,368 tons of peanuts and 6,958 tons of rapeseeds. Among the oil-bearing crops, sesame has the highest oil content. Other important crops for food manufacturing include sugar-cane, beetroots, tea, tobacco and a variety of fruits. While the production of these crops more than doubled during 1978-1990, spectacular production increases were recorded by beetroots, citrus, grapes and bananas.

The livestock and fishery resource base for food manufacturing finds its expressions in data presented in Tables IV.2 and IV.3. China ranks first in the global poultry and egg output, and second in meat production. China is also the world's second largest aquatic products producer, with a record output of 12.4 million tons in 1990.

The primitive mode of supplying fodder is a serious constraint inhibiting the rapid progress of animal husbandry in China. During the second half of 1988 and throughout 1989 the dairy industry in China experienced a shaky period when farmers could not afford to feed their cattle. Prices for fodder tripled against a 50 per cent rise in the retail prices for liquid milk. The situation improved in 1990, when the provincial governments agreed that in exchange for delivery of a specified quantity of milk, farmers could buy fodder at subsidized State prices. A number of foreign aid programmes, mostly from Scandinavian countries, are under way to help Chinese cope with growing demand for dairy products.

Total fishery production, including seawater and freshwater varieties, stood at 12.4 million tons in 1990. Although there has been a rapid growth in fisheries production, there is still much room for further expansion. There are 14 million hectares of shallow sea and beaches, yet less than one-seventh is used for aquatic breeding. Of some 20 million hectares of inland waters, only one-fifth is used, with low yield. The recently started deep-sea fishing has opened up new avenues for the development of the country's aquatic products industry.

Key constraints are lack of funds, technical know-how, unattractive returns and high fuel costs. The government is encouraging paddy field and beach-yard fish farming. Paddy field culture relies on the natural nutrients in the water to sustain the fish whose excreta, in turn, raise soil fertility. Around 600,000 hectares of rice fields are used for fish breeding. Introduction of contract fish farming has contributed to a significant increase in yields.

The contours of the Eighth Five-Year Plan (1991-1995) and the Ten-Year Development Strategy (1991-2000) spell out specific measures to boost agricultural production. The proposals include rural reforms, linking output to remuneration, improving the dual management system, and a different form of optimum-scale farming befitting local conditions. The proposals also aim at constructing a number of medium- and large-size flood control, water storage and water diversion projects. Efforts will also be made to build a number of major agricultural commodity centres. The greater part of the increased agricultural output is envisaged to result from the popularization and application of scientific and technological achievements in the 1990s.

The proposals are also aimed at reforming the distribution system of farm produce in order to facilitate, among other things, processing facilities. The contours of the Eighth Plan and the Ten-Year Development Strategy also call for efforts to guide rural industries in adjusting their structure in order to improve product quality and to sustain a healthy pace of expansion in the 1990s.

Table IV.1. Output of major food crops, 1978-1990

Crops	1978	1979	1980	1981	1982	1983	1984	1985	1986	1987	1988	1989	1990
Grains (million tons)													
Rice	136.93	143.74	139.91	143.96	161.60	168.87	178.26	168.57	172.22	174.26	169.11	180.13	189.33
Wheat	53.84	62.73	55.21	59.64	68.47	81.39	87.82	85.81	90.04	85.90	85.43	90.81	98.23
Corn	55.95	60.04	62.60	59.21	60.56	68.21	73.41	63.88	70.86	79.24	77.35	78.93	96.82
Soybeans	7.57	7.46	7.94	9.33	9.03	9.76	9.70	10.50	11.61	12.47	11.65	10.23	11.00
Tubers	31.74	28.46	28.73	25.97	27.05	29.25	28.48	26.04	25.34	28.20	26.97	27.30	27.43
Oil-bearing crops (1,000 tons)													
Peanuts	2,377	2,822	3,600	3,826	3,916	3,951	4,815	6,664	5,882	6,171	5,693	5,363	6,368
Rape seeds	1,868	2,402	2,384	4,065	5,656	4,287	4,205	5,607	5,881	6,605	5,044	5,436	6,958
Sesame	322	417	259	510	342	349	476	691	618	526	404	338	469
Sugar-cane (1,000 tons)	21,116	21,508	22,807	29,668	36,882	31,141	39,519	51,549	50,219	47,363	49,064	48,795	57,620
Beetroots (1,000 tons)	2,702	3,106	6,305	6,360	6,712	9,182	8,284	8,919	8,306	8,140	12,810	9,243	14,525
Tea (1,000 tons)	268	277	304	343	397	401	414	432	461	508	545	535	540
Flue-cured tobacco (1,000 tons)	1,052	806	717	1,279	1,848	1,151	1,543	2,075	1,374	1,636	2,337	2,405	2,259
Fruits (1,000 tons)													
Apples	2,275	2,869	2,363	3,006	2,430	3,541	2,941	3,614	3,337	4,264	4,344	4,499	4,319
Citrus	383	555	713	798	939	1,296	1,499	1,808	2,548	3,224	2,560	4,561	4,855
Pears	1,517	1,438	1,466	1,593	1,755	1,795	2,100	2,137	2,348	2,489	2,721	2,565	2,353
Grapes	104	126	110	148	186	247	294	361	442	641	792	874	859
Bananas	85	74	61	126	201	207	300	631	1,251	2,029	1,830	1,404	1,446

Source:State Statistical Bureau of China, *China Statistical Yearbook* (various issues).

Table IV.2. Output of livestock products, 1987-1990
('000 tons)

	1987	1988	Change of 1988 over 1987		1989	1990	Change of 1990 over 1989	
			Quantity	Percentage			Quantity	Percentage
Pork, beef and mutton	19,860	21,936	2,076	10.5	23,262	25,135	1,873	8.1
Pork	18,349	20,176	1,827	10.0	21,228	22,811	1,583	7.5
Beef	792	958	166	21.0	1,072	1,256	184	17.2
Mutton	719	802	83	11.5	962	1,068	106	11.0
Cow's milk	3,301	3,660	359	10.9	3,813	4,157	344	9.0
Sheep and goat's milk	487	529	42	8.6	545	594	49	9.0
Honey	204	195	9	-4.4	189	193	4	2.1
Poultry eggs	5,902	6,955	1,053	17.8	7,198	7,946	748	10.4

Source: State Statistical Bureau of China, *China Statistical Yearbook* (various issues).

Table IV.3. Output of aquatic products, 1987-1990
('000 tons)

Item	1987	1988	1988 as percentage of 1987	1989	1990	1990 as percentage of 1989
Total	9,553	10,609	111.1	11,517	12,370	107.4
Seawater aquatic products	5,481	6,057	110.5	6,612	7,133	107.9
Naturally grown	4,381	4,633	105.8	5,036	5,509	109.4
Artificially cultured	1,100	1,424	129.5	1,576	1,624	103.0
By category						
Fish	3,516	3,629	103.2	3,830	4,223	110.3
Shrimps,prawns and crabs	840	1,033	123.0	1,057	1,070	101.2
Shellfish	889	1,144	128.7	1,375	1,473	107.1
Algae	236	251	106.4	300	275	91.7
Main seawater aquatic products						
Greater croaker	17	18	105.9	19	25	131.6
Lesser croaker	20	24	120.0	17	23	135.3
Hair tail fish	393	366	93.1	416	498	119.7
Scad and mackerel	511	241	47.2	232	585	252.2
Sea tangle	179	216	120.7	273	244	89.4
Freshwater aquatic products	4,072	4,552	111.8	4,905	5,237	106.8
Naturally grown	600	654	109.0	735	1067	145.2
Artificially cultured	3,438	3,898	112.3	4,170	4,170	100.0
By category						
Fish	3,938	4,408	111.9	4,728	5,049	106.8
Shrimps, prawns and crabs	81	85	104.9	95	95	100.0
Shellfish	53	59	111.3	73	76	104.1

Source: State Statistical Bureau of China, *China Statistical Yearbook* (various issues).

Emerging trends

In 1988, grain and edible oil processing accounted for around 25.2 per cent of the food industry output. The second largest segment of the food industry was cigarette manufacturing, which accounted for 20.6 per cent of output in 1988, the latest year for which disaggregate data are available.[2/] Slaughtering and meat processing contributed around 8.4 per cent to the output of the Chinese food industry in 1989, and beverages and bakery products accounted for 17.3 per cent and 7.4 per cent, respectively. Thus, the major segments of food manufacturing in China include grain and edible oil processing, cigarette manufacturing, meat processing, beer and wine production, and bread and bakery products.

In 1988 food manufacturing in China encompassed 42,755 enterprises and generated output worth Rmb 73,071 million in constant 1980 prices. The food industry in China is relatively regionally concentrated. The six leading provinces - Jiangsu, Sichuan, Guangdong, Shandong, Heilongjiang and Liaoning - accounted for over 45 per cent of the gross value of industrial output in 1988, compared with 37 per cent in 1982. On the other hand, Guizhou, Tibet, Gansu, Qinghai and Ningxia produced around 5 per cent of gross output value in 1988. The second half of the reform decade witnessed a modest increase in industrial concentration.

Whereas the growth of the traditional products - sugar, salt - has been modest and uneven, canned food and beverage production has tripled. Cigarette production has also increased significantly (see Annex Table A-14). This shows the shift towards more expensive food products. Within the canned food industry particularly rapidly expanding items are poultry, meat and fruit products. Vegetable foods have registered a decline over the period 1983-1988. Dairy products have grown rapidly. Milk products, confectioneries and snack food are strong growth areas. Nevertheless, less than a third of the food produced is processed, and wastage (especially of fruits and vegetables) is very high.

Fig.IV.1. Physical output of selected food products, 1980 and 1990 (Million tons)

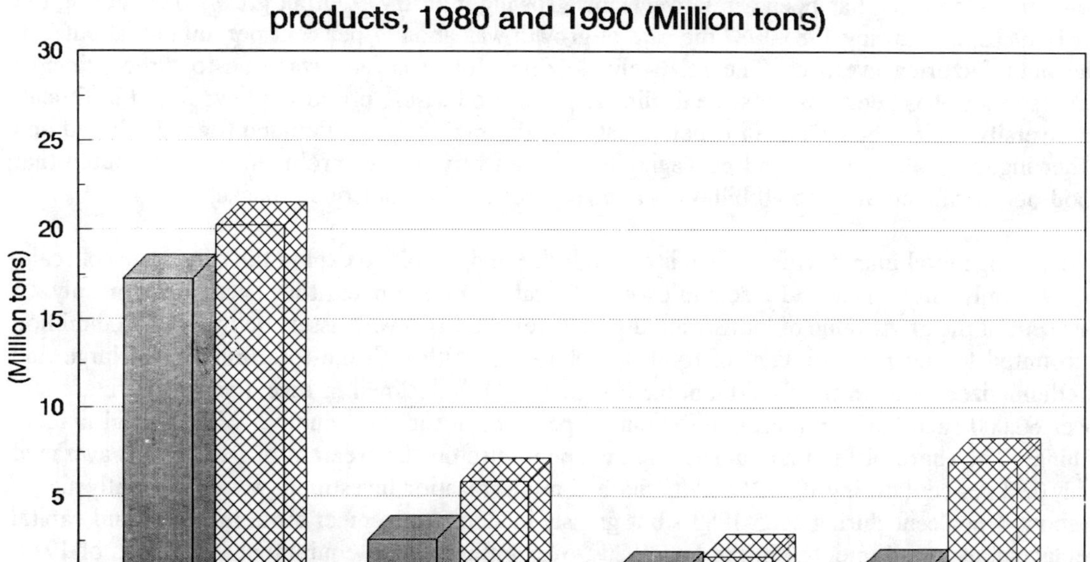

China's beverage industry has grown rapidly in recent years. The production level of 12.8 million tons of beverages in 1989 represented 4.2 times more than the 1978 level. A number of newly developed soft drinks, including fruit and vegetable juice and mineral water have achieved high quality standards.

Food exports

Aquatic products rank first in China's food exports in dollar terms, followed by canned food, tea and meat products (see Table IV.4). The Chinese tea industry has long held a special place in the country's export profile. Export earnings from tea totalled $412.7 million in 1990. China is the world's third largest exporter of tea after India and Sri Lanka. More than 80 countries buy Chinese tea. Although average yields are low, exports grew significantly in 1988 and 1989. China currently produces six main varieties of tea, with different types of processing. However, varieties of tea are named after the colour of infusion, i.e., green, yellow, black, white, grey and red, rather than that of the processed leaf. All these are unrefined and are subject to grading, blending and packaging before final sale. China also produces scented and compressed teas from either crude or graded teas. The delicate flavour of a popular Chinese tea - oolong tea - is now more readily available for project development specialists, who formulate their own distinctive tea-based beverages.[3/]

Among the canned food products, canned vegetables constitute a major export item, earning over $360 million in 1990. Export earnings from canned fruits were to the tune of $53.3 million in 1990. Most of the canned fruits and vegetables are fruits in syrup, jam and vegetables with spicy water. Major canned fruits include mushrooms, citrus, pineapples, asparagus, water chestnuts, green beans, peas, bamboo sprouts, tomatoes, broad beans, apples, pears and peaches. The dried Chinese dates and jujubes, slices of bamboo sprouts, yellow day lilies, mushrooms, pickling and salting vegetables have been traditionally famous Chinese products which are prominent on the world market.

Problems and prospects

Food manufacturing has been a relatively slow growing industry. Output grew slowly during the early 1980s, and during 1985-1987 the rate of growth was about 7 per cent per annum, about half the manufacturing average. The relatively slow pace of expansion was due to three principal causes: there has been a large-scale decline in grain production; producers have found it difficult to diversify their product mix in order to cater to the rapidly rising demand for meat-based and other higher quality foods; and packaging has been a very serious problem. It is estimated that food output valued at Rmb 40 billion is annually lost due to inadequate packaging.

The average food manufacturing firm is of small size and unable to exploit the economies of scale. In 1987 only three firms had assets in excess of Rmb 100 million and they accounted for only 2.5 per cent of the gross value of industrial output. Over 600 firms with assets below Rmb 300 million accounted for over 70 per cent of total output value. Although the total number of large and medium-sized firms in the food manufacturing sector has declined in recent years, there are still over 60,000 such firms, representing about 13 per cent of the total number of industrial units in China. The share of food manufacturing in expenditure on the creation of fixed assets averaged only 0.93 per cent during 1985-1988. Its share in modernization investment was significantly higher - about 4 per cent during 1985-1988 - but grossly inadequate to meet modernization and capital replacement requirements. In the late 1980s, over a third of the equipment in use was of 1940s vintage. Nevertheless, some progress has been made in achieving modernization in the larger firms. The most successful enterprises in achieving quality upgrading include the Chongqing Tinned Food Factory, the Jilin Daan Canning Factory, the Yantai Grape Wine Company, the Guangzhou Beer Brewery, and the Ninchang Egg Factory. In recent years, the food industry has also benefited from increasing imports of foreign technology, and adaptation of successful production and marketing methods.

Table IV.4. Volume and value of manufactured food exports, 1987-1990
(Value in million $)

Item	Unit Volume	1987 Volume	1987 Value	1988 Volume	1988 Value	1989 Volume	1989 Value	1990 Volume	1990 Value
Beef, fresh and frozen	Tons	33589	55.32	53986	107.98	56494	105.94	96593	158.74
Pork, fresh and frozen	Tons	99964	172.08	63484	115.82	88424	159.30	124236	215.48
Frozen chicken	Tons	16768	25.23	25660	43.48	31466	55.31	37813	74.39
Rabbit meat frozen	Tons	20545	40.91	20976	37.39	21438	34.61	17520	30.08
Aquatic products	Tons	218210	702.59	287229	932.85	293916	991.60	357544	1316.10
Dried mushroom	Tons	2640	22.56	6004	48.97	5413	39.61	6698	47.93
Edible fungus	Tons	1561	11.85	1957	14.77	2356	14.57	1860	10.00
Hot pickled mustard tuber	Tons	14134	5.06	12584	4.72	13940	4.68	12911	4.43
Bitter apricot kernel	Tons	7934	9.75	7452	8.54	7745	9.85	8180	9.39
Walnut meat	Tons	11293	24.87	10608	23.51	12846	27.16	8712	18.50
Sugar	Tons	452494	90.56	247802	62.04	429623	161.41	570493	229.81
Honey	Tons	66831	53.75	46487	37.02	71499	56.14	88005	71.71
Tea	Tons	174279	362.49	198360	401.94	204584	420.74	195471	412.71
Dried chilies	Tons	33232	24.42	32206	29.69	39864	49.21	40722	37.60
Canned food	Tons	536974	535.95	554210	649.04	548357	674.28	565748	681.41
Canned pork	Tons	93757	156.42	81558	143.32	86340	147.54	90906	152.88
Canned vegetables	Tons	329858	282.09	333228	353.10	332145	366.56	332709	361.31
Canned fruits	Tons	87351	59.94	87967	58.74	71399	51.71	77825	53.25
Beer	Tons	32428	13.58	39343	20.64	41752	25.18	35223	19.57
Flue-cured tobacco	Tons	17019	35.50	19367	41.49	21931	47.66	28735	49.36
Salt	Million tons	1.7	24.76	0.38	14.54	0.37	15.63	0.48	18.23
Edible vegetable oil	Tons	55659	30.93	25503	17.37	62099	39.45	139477	93.42

Source: State Statistical Bureau of China, *China Statistical Yearbook* (various issues).

The vertical and horizontal integration of the food processing industry has progressed very slowly. The food processing industry produces a large number of products in widely dispersed units with management of individual enterprises having little direct contact with their counterparts. There exists a significant level of functional duplication and executive overlapping. Horizontal integration and production rationalization is hampered by provincial and local concern to achieve and maintain food self-sufficiency. A weak transportation infrastructure discourages production rationalization. Hence, Beijing, Tianjin and Shanghai all aimed to become self-sufficient in grain production by 1990.[4/] Perishable vegetables, fruits and fish are concentrated in city suburbs due to inadequate storage and refrigerating facilities. The integration of storage, transportation, processing and marketing units is at a preliminary stage. Investment would need to be made to achieve such integration as well as to provide technical support to improve the quality of agricultural supplies which is at present low - there is a scarcity of high protein flour needed to make high-grade bread and lean meat necessary for the production of better quality processed fast foods. Processing raw materials are also of variable quality - the dirt and moisture content of oilseeds is high.

Contractual agreements between processors and peasants for vegetables have increased. An annual co-ordinating conference is held to supervise contracts over raw material supplies. Integrated systems for breeding, planting and processing are gradually emerging in many product areas - notably citrus fruit - with the encouragement of the Ministry of Agriculture.

Research and Development (R & D) within the food manufacturing sector remains low. An important concern of food research centres at Beijing, Guangdong and Sichuan is the development of the food packaging industry - packaging technology which lengthens the life of unrefrigerated foods. Small-scale production of food plastic, waxed cartons and thermo forms is under way. A joint venture with the Swedish firm Tetrapak has been established for the development of paper packaging.

The development of the packaging industry is a prerequisite for the rapid growth of food manufactures exports. This requires an expansion of the domestic production of quality paper, plastics and glass and also a standardization of the production procedures for cans and bottles. Paper quality can be improved by substituting wood pulp for plant fibre. Canning techniques - involving better soldering methods - also need to be upgraded rapidly. China attaches high priority to the packaging industry and there is considerable potential for international companies that can assist China - through both licensing and joint venture arrangement - in expanding and upgrading the domestic production capacity in the food packaging branch. The focus of development in food packaging is on sterilization technology, materials and processes, de-oxygenation packaging, inflated packaging for processing and preserving freshness in meat, eggs, vegetables and fruit, and various kinds of composite materials, light bottles, punched double-piece cans and high-frequency welded double-piece cans needed in the packaging of drinks, condiments and edible oils.

It is estimated that, by the year 2000, the national consumption of meat, poultry, eggs and fish will have doubled, and the projected grain output will be far below demand. In the face of the proportion of animal food failing to keep pace with increasing demand, it is imperative to gradually develop a food consumption structure suited to China's actual conditions and to make the food industry well-matched with the structure. In order to achieve this, priorities are accorded to selected segments of the food industry. In cereal and oil processing, the rice- and flour-producing segments are destined for expansion. The quantity and varieties of wheat flour will also be increased. Oil-refining, purifying, decolouring and deacidifying processing will be upgraded. Special oils for food, such as hydrogenated oils, margarine and shortening, are to be developed. In sugar production efforts are under way to develop large-scale manufacture of products based on beet and cane sugar. Attention is also focused on combining coarse grains with refined grains, to produce foods of a reasonable nutritional value appropriate to China's resource endowment and pattern of food consumption. The scope for identifying food products which have an export potential would need to be explored. They could carefully be targeted to specific markets and a range of policies could be designed to achieve penetration within these markets.

B. TEXTILES AND CLOTHING: ACCENT ON UPGRADING

The resource base

The main fibre crops constituting the raw material resource base for the Chinese textile industry are cotton, hemp crops for jute and linen manufacturing, and silk. China's five main cotton areas include the Chang Jiang valley, the Huang He valley, and north, north-west, and south China. The most important area is the Chang Jiang valley, which has the highest yield per unit of area producing the largest quantity of cotton. The area under cotton cultivation rose significantly in 1988, but cotton output fell to 4.15 million tons, from 4.25 million tons in 1987 (see Table IV.5) as a result of an 8.6 per cent decline in yield. In 1990, both sown area and output rose significantly. A rise in cotton output in 1990 was atributed largely to a significant increase in purchasing prices, favourable weather conditions, and to a more widespread application of scientific and technological methods in cotton production.

Since the mid-1980s China has emerged as the world's largest cotton producer and a major exporter. While domestic demand has been rising, cotton production has been falling from a record harvest of 6.2 million tons in 1984, due to negligent State pricing policies and the tendencies of farmers to shift to more profitable cash crops. According to an official estimate, China needs at least 4.5 million tons of cotton to cope with the domestic industrial demand. The cotton output in 1990 was significantly higher than the 1989 output of around 3.8 million tons. However, cotton scarcity is exacerbated by falling stockpiles accumulated in the early 1980s.[5/] If the supply situation does not improve China may have to begin importing cotton again as it did in the late 1970s.[6/] The total demand for cotton on the Chinese market stood at around 5 million tons in 1990.

China produces a wide variety of hemp crops (see Table IV.5). Jute is produced mainly in Zhejiang, Guangdong, Hunan, Jiangxi and Jiangsu provinces in the Chang Jiang and Zhu Jiang basins. Ramie is grown only in China, mainly in Hunan, Hubei and Sichuan provinces. Cloth woven from ramie and polyester fibre made in China is popular on the world market. Linen fibre made of flax crop is suitable particularly for waterproof cloth. With the exception of flax crop, the output of other hemp crops fell significantly in 1988. In 1989 and 1990, the output of flax crops fell significantly.

Table IV.5. Sown area and output of cotton, hemp crops and silkworm cocoons, 1987-1990

	Sown area (1,000,0000 mu)[a/]				Output (In thousand tons)			
Item	1987	1988	1989	1990	1987	1988	1989	1990
Cotton	72663	83021	78050	83822	4245	4149	3788	4508
Hemp crops	145000	11031	8450	7427	2084	1809	1124	1097
Jute and ambary hemp	4082	4158	4288	4498	1138	1078	660	726
Ramie	7747	4076	2233	1208	568	318	184	89
Flax	1891	2203	1453	1306	320	369	243	241
Silkworm cocoons	18125	17860	18007	19367	402	441	488	535
Mulberry silkworm cocoons	5065	5172	5747	7261	354	394	435	480
Tussah silkworm cocoons	13060	12688	12260	12106	49	46	53	54

Source: State Statistical Bureau of China, *China Statistical Yearbook* (various issues).

a/ 1 mu = 1/15 hectare.

China ranks first in the production of mulberry-fed silk and tussah silk cocoons. The country's output of silk accounts for one-third of the world total. The output of mulberry silkworm cocoons rose significantly in 1988, while that of tussah silkworm cocoons suffered a marginal decline. According to rough estimates China produces around 90 per cent of the world's tussah cocoons output.

Other sources of raw materials for the textile industry include sheep wool (fine wool), goat wool and cashmere. The output of these items in 1988 stood at 110,676 tons, 14,425 tons and 4,710 tons respectively. The production capacity of wool textile industries has been expanding through modernization and introduction of new varieties.

Emerging trends

China's textile and clothing industry has made significant headway in production and exports. Despite raw material shortage, energy and domestic demand constraints, the gross output value maintained a modest increase of 3 per cent, while the export value of textiles and garments rose by over 10 per cent in 1989. Two-thirds of textiles and garments are sold in the domestic market, and one-third is destined for exports, making the country one of the largest exporters in the world. China ranks first in cotton yarn export and sixth in garment export. China's exports of woollen fabrics, woollen blankets, cashmere sweaters, knitted outerwear, bed sheets and chemical fibre textiles take a significant share in world exports of these items. One-quarter of the country's total foreign exchange stems from textile and garment exports.

Currently there are over 25,000 enterprises producing textiles. Around 22,000 enterprises produce items of clothing. With 31.54 million cotton spindles, China has the biggest production capacity for cotton yarn and cloth in the world. The country is the second biggest wool producer with around 2.26 million wool spindles. Collective enterprises are overwhelmingly important in the clothing sector, accounting for almost 85 per cent of all employment within the clothing sector. However, collective enterprises provide only about 25 per cent of textile employment.

The textile and clothing industry is heavily concentrated in Jiangsu, Zhejiang, Shanghai and Shandong. Together these four provinces accounted for roughly half of the textile sectors gross output during the late 1980s. As against this, the corresponding share of Tibet, Ningxia, Qinghai and Guizhou was only 0.6 per cent in textiles and 1.1 per cent in clothing.

Physical output indices of major textile and clothing items show a steady and accelerated expansion of major items during 1980-1990 (see Annex Table A-15). However, against a significant increase in output the industry's performance has relatively slowed down since 1981. During the period 1979-1981 the textile industry was operating in a seller's market. From 1982 onwards, demand has slackened somewhat and - what is more significant - a divergence has appeared between the product mix supplied and the composition of market demand. Textile products have increasingly failed to meet quality, variety, colour and style requirements. Beginning in 1982, the textile industry started to produce a growing unsaleable surplus. Readjustment of production programmes requires modernization and technical adjustment in production equipment. A large proportion of current equipment is not usable for the efficient production of higher quality products. The pressing need for equipment modernization is illustrated by the trend of declining productivity in this industry for most of the 1980s.

The net output to gross output ratio (roughly equivalent to value added as a proportion of gross output) equalled 23 per cent for textiles and 27 per cent for clothing in the late 1980s, compared to a national industrial average of 31 per cent.[7/] Labour productivity declined by 3.1 per cent in textiles and 29 per cent in clothing in 1987, compared to 1985. The declining trend in productivity particularly in textiles persisted until 1988.

The textile industry registered a 4.9 per cent real increase in labour productivity in 1989, while the apparel segment recorded an 18.9 per cent increase in labour productivity in the same year. The exceptionally high growth of gross output value per worker in 1989 could be attributed to a fall in the number of workers. The government tried to introduce effective measures towards readjusting product structure and increasing the number of varieties to meet the changing pattern of consumption. The year 1990 was named as the "year of variety and quality" by the Ministry of Textiles with a view to injecting a fresh impetus towards enhancing the quality of textile products. In order to alleviate the shortage of raw materials, efforts are under way to raise the proportion of chemical fibres and the use of non-cotton fibres, such as blend fabric and interwoven products.

Despite the rapid growth of synthetics and blends, improvements in quality cloths and weaving apparel are obstructed by the limited progress achieved in production rationalization and horizontal integration. The average size of the textile enterprises remains small - there are over 50,000 town and village enterprises which produced a fifth of all clothing in the country in 1988. The growth of subcontracting links and the formation of producer and trading associations did not significantly rectify structural weaknesses and inefficiencies. Production structures within the major textile producing provinces are similar. There is very little product specialization among the provinces. Moreover, there is no evidence that the level of provincial specialization has increased in recent years.

Cotton shortage and idle capacity

The extent to which cotton is needed to keep the installed spindles and looms running can be gauged from coastal textile enterprises' situation as depicted in Table IV.6. Allocated cotton is short of full-capacity requirements, and actual delivery is far from planned allocation. In the industrial heartland of Shanghai actual delivery of cotton was around 36 per cent of the planned allotment and 26 per cent of the amount needed, leading to substantial underutilization of cotton spinning capacity and 7,000 persons out of work in the second half of 1990. This situation can be generalized to textile enterprises located in 12 Chinese coastal cities. Lack of working capital, cash flow overdue and soaring costs of raw materials altogether have led to a severe burden on the textile industry in China.

Fig.IV.2. Physical output indices of major textile and clothing items, 1980 - 1990 (1980 = 100)

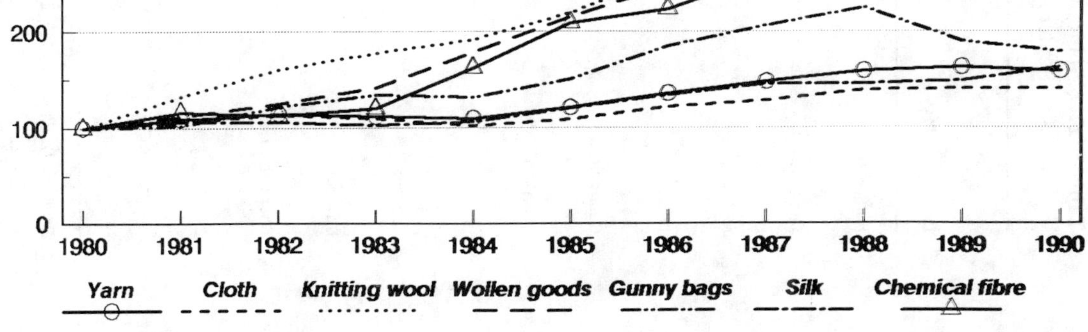

Table IV.6. Spinning capacity, cotton supply and work stoppage in textile enterprises located in twelve coastal cities, September 1989 - March 1990

Coastal city	Cotton spinning capacity		Cotton supply September 1989-March1990 (tons)			Work stoppage as of 31/3/1990 Workers	
	Total spindleage	Idle capacity	Needed	Allocated	Delivered	Employed	Not working
Hangzhou	230,000 spindles 3,200 rotors	20,000 1,000	4,000	1,000	600	9,000	1,700
Shanghai	2,300,000 spindles 15,000 rotors	80,000	30,000	22,400	8,000	61,160	7,000
Nantong	760,000 spindles 15,600 rotors	 5,600	12,000	5,700	2,850	..	..
Foshan	185,000 spindles 4,944 rotors	20,000	3,750	650	210	2,500	300
Guangzhou	246,000 spindles 7,800 rotors	100,000 7,800	4,300	1,620	550	6,100	2,500
Tianjin	780,000 spindles 5,000 rotors	200,000	11,000	7,150	2,900	50,000	10,000
Wuxi	410,000 spindles 6,600 rotors	 2,000	6,200	3,280	2,100	..	..
Suzhou	450,000 spindles 3,600 rotors	 1,000	6,000	3,460	2,100	..	..
Dalian	400,000 spindles 4,000 rotors	..	6,000	2,500	750	12,000	..
Changzhou	340,000 spindles 16,672 rotors	52,000	6,000	3,590	2,175	10,800	..
Beijing	499,000 spindles 4,000 rotors	50,000 1,500	7,500	6,000	2,550	14,000	3,000
Qingdao	484,000 spindles 10,600 rotors	30,000 2,500	7,250	6,300	3,100	35,000	2,400

Source: *Textile Asia* (July 1990), p. 93.

As the short supply of cotton has seriously disrupted textile production, China has had to depend on imported supplies at high prices, leading to higher production costs and loss of competitiveness. Increased domestic production and efficient State purchase without distortions on the price front could pave the way for a less chaotic situation in the 1990s.

Man-made fibre to the fore

The growth of man-made fibre production has outpaced that of cotton production. Annual production of man-made fibre was estimated to reach 1.5 million tons in 1990, replacing a substantial quantity of imported man-made fibre. Imports of man-made fibre were expected to fall from 600,000 tons in 1989 to 500,000 tons in 1990. The development of man-made fibre remains a focal point in the textile industry development strategy, with new types of polyester fibre, acrylic fibre and polyamide fibre as important segments. In the face of the rapidly increasing production of polyester filament in the 1980s, the production of synthetic filament was around 20 times higher than in the late 1970s. The total amount of man-made fibre required by the Chinese textile industry in 1990 was estimated at 1.9 million tons.

The production of man-made fibre received a significant boost when the first plant of Yizheng Joint Corporation of Chemical Fibre Industry (YJCCFI) became fully operational by the end of 1989. This giant fibre production centre is located on the northern bank of the Yangtze River, 70 km east of Nanjing, the capital of Jiangsu province. With a production capacity of 240,000 tons of staple fibre and 260,000 tons of polyester chips, it is capable of producing 500,000 tons of polyester. China used to import around 200,000 tons of polyester a year in the 1980s. The country is still importing some polyester fibres and chips to meet domestic demand. With an accelerated production expansion, man-made fibre is expected to make up more than 30 per cent of all fibres produced in China by the year 2000.[8/]

Silk situation warrants innovation

China aims at increasing the annual value of exports of silk textiles from $1,650 million in 1988 to $2,500 million in the first half of the 1990s. This ambitious target calls for the use of modern technology to improve product lines, and the use of efficient modes of production. The silk production target set for 1990 was 1,000 million tons less than the production level achieved in 1988. The production target of 1,530 million metres of silk fabrics for 1990 was also less than the target of 1,711 million metres in 1988.

As the output of raw materials is steadily on the decline, together with the quality of products, new difficulties confront the silk manufacturing enterprises. China's silk textiles on the world market command lower prices than similar items from the Republic of Korea and Japan. There is an urgent need to adopt technological renovation in order to improve product quality and to develop varieties to meet the changing pattern of demand on the world market. For a long time the designs of silk products have remained old-fashioned. With more diversified silk products even domestic sales could be significantly increased to the advantage of the industry.

Priority areas

In the sphere of utilizing foreign investment and advanced technology priority is accorded to the following:

- developing the chemical fibre industry in order to increase varieties and improve quality through the establishment of huge acrylic fibre mills and viscose acetal fibre mills;

- finishing treatment process in dyeing and pure cotton fabrics, terylene blends, woollen imitation fabrics, pure cotton knitwear, terylene filaments and shuttle woven fabrics in order to keep pace with the changing patterns;

- advanced textile machinery and testing instruments used in the technical transformation of existing enterprises;
- product design improvements in medium- and high-grade garments;
- developing new varieties of raw materials and production bases for raw materials; and
- setting up external sales and information networks.

Textile and clothing export prospects

In the first half of 1989 Chinese exports of textiles and clothing fetched $5,443 million (see Table IV.7), accounting for 24.5 per cent of total exports. In dollar terms significant export items include women's woven coats, pure cotton cloth, men's woven coats, pure silk fabrics, knitted underwear, knitted coats, woven underwear, poly-cotton cloth, pure cotton yarn, pure wool carpet, table cloths, bed sheets, ramie and linen cloth, flax and ramie blended yarn and cotton bath towels.

Table IV.7. China's textile and clothing exports, January-June 1989

Item	Quantity	$'000
Silk yarn	2,052 MT	61,390
Woollen yarn	6,562 MT	80,620
Pure cotton yarn	479,000 bales	207,400
Cotton thread		5,420
Synthetic yarn	59,700 bales	27,500
Poly-cotton yarn	132,400 bales	57,770
Flax and ramie blended yarn	221,400 bales	98,160
Man-made fibre yarn	55,700 bales	27,550
Pure cotton cloth	979,110,000 metres	665,240
Poly-cotton cloth	464,500,000 metres	275,420
Artificial silk fabrics	4,990,000 metres	7,500
Pure silk fabrics	88,930,000 metres	335,870
Gunny bags	45,000,000 pieces	15,880
Woollen blankets	257,000 pieces	8,200
Woollen and blended items	5,049,000 metres	35,530
Ramie and linen cloth	75,380,000 metres	99,280
Bed sheets		106,010
Table cloths		123,490
Cotton bath towels	63,250,000 pieces	91,540
Cotton washing towels	100,810,000 pieces	21,770
Square towels	203,860,000 pieces	18,420
Other cotton towels	171,590,000 pieces	28,520
Pure wool carpet	2,310,000 square metres	170,880
Synthetic carpets	1,000,000 square metres	3,470
Men's woven coats	104,880,000 pieces	426,080
Women's woven coats	196,770,000 pieces	714,610
Woven underwear	144,200,000 pieces	276,200
Knitted coats	105,910,000 pieces	277,160
Knitted underwear		317,550
Handkerchiefs	174,280,000 pieces	27,330
Gloves and stockings	253,280,000 pairs	66,840
Total, including other		**5,443,620**

Source: *Textile Asia* (October 1989), p. 77.

China's textile and clothing export target for the year 2000 stands at $20 billion. Hong Kong, Taiwan Province and the Republic of Korea are the potential competitors. The Republic of Korea's textile and clothing export target for 2000 is set at $30 billion. In 1970, textile and clothing exports of Taiwan Province and the Republic of Korea were only half those of China. By 1987, the Republic of Korea's textile and clothing exports exceeded China's by $2 billion, while Taiwan Province emerged as the world's sixth largest exporter, lagging slightly behind China. In 1970, Hong Kong's textile and clothing exports were almost equal to those of China. By 1980, Hong Kong outpaced China in textile and clothing exports, and emerged as the world's largest textile exporter. Although China still enjoys the advantages of relatively low labour cost, technological proficiency in China is lagging behind the east Asian Newly Industrializing Economies whose textile products have a strong position on the world market.

Hong Kong, the United States, Japan, the EC and Canada are the major export destinations for China's textiles and clothing, absorbing around 70 per cent of the total export. Almost half of all Chinese textile products are destined for Hong Kong. The scale of Hong Kong's dependence on China for manufacturing activities is revealed by Hong Kong's trade performance in the first half of 1990. During this period, for every dollar worth of domestic exports to the world market, Hong Kong shipped around two dollars worth of re-exported goods from China.

Apart from Hong Kong and Japan, the major target markets continue to be the United States and the EC. The Multi-Fibre Arrangement (MFA) restrictions do not seem to have prevented the rapid growth of the Chinese textile exports in general to the United States and the EC. This was partly due to the price competitiveness of Chinese non-quota items. A significant increase in the share of clothing in China's export profile during 1978-1988 partly reflected a shift in commodity composition to non-restricted items.

Table IV.8. Quotas on China's textile exports to the United States, 1989

Category	Description	Quota
200	Sewing thread/yarn, retail quality	1,164,000 pound
218	Coloured woven cloth fabrics, cotton or man-made	10,736,432 square metres
219	Canvass, cotton or man-made	2,280,935 square metres
300/301	Cotton yarn	7,175,000 pound
239	Baby wear, cotton or man-made	3,893,400 pound
313	Woven cloth	52,872,723 square metres
314	Shandong pongee	29,755,512 square metres
315	Cloth, for printing	179,500,000 square metres
333	Men's and boys' jackets	70,350 dozen
334	Other types of men's and boys' coats	253,448 dozen
335	Women's and girls' coats	333,775 dozen
336	Skirt with blouses	133,900 dozen
337	Casual and sports wear	1,111,950 dozen
338/339	Shirts, knitted	2,045,160 dozen
340	Shirts, woven for men and boys	734,514 dozen
341	Shirts, woven for women and girls	575,680 dozen
342	Short skirts	109,200 dozen
347/348	Trousers, shorts and casual wear pants	2,127,980 dozen
350	Dressing gowns	119,700 dozen
351	Pyjamas	388,500 dozen
352	Underwear	1,495,395 dozen
359-C	Work cloth pants	964,160 pound
359-V	Vests	1,534,700 pound
360	Pillow cases	4,099,400 pieces
361	Bed sheets	3,382,560 pieces
363	Towels	25,357,500 pieces
369-D	Napkins	8,772,000 pound
443	Suits for men and boys	127,260 sets
444	Suits for women and girls	189,193 sets
445/446	Woollen knitwear	272,700 dozen
607	Short synthetic yarn	5,665,000 pound
631	Gloves and boxing gloves	910,011 dozen pairs

Source: *Textile Asia* (June 1989), p. 61.

Quotas on China's textile exports to the United States are presented in Table IV.8. China has consistently made good use of quotas. China's utilization rate of the quota imposed by the United States exceeded 100 per cent in the early 1980s, and its average quota utilization level for Canada stood at 98.3 per cent during 1980-1986. China was also able to make full utilization of its quota in the EC. Full utilization of quota indicates that China's exports are artificially restricted by quantitative limits imposed on them. China fully utilizes quotas for most restricted items as do Newly Industrializing Economies that are competitive on the strength of high productivity levels, quality, reliability and good marketing and distribution channels.

China's textile products have gained increased access to the EC in recent years. During 1984-1987 MFA products nearly doubled in volume, and rose by over 80 per cent in European currency unit terms. This was due largely to quota overruns in 1986 and 1987. China's exports of selected items to the EC in 1987 was higher than quantitative limits imposed on those items in 1988 and 1989 (see Table IV.9). Quota on T-shirts was significantly increased from 6.6 million pieces to 32.8 million pieces, while quantitative limits were imposed on gloves and mittens for the first time in 1989.

Table IV.9. Chinese textile exports to the EC and quantitative limits, 1987-1989

EC category		Unit	Exports 1987	Quantitative limits 1988	1989
1	Cotton yarn	tons	3,924	2,973	3,300
2	Cotton fabrics	tons	38,404	21,013	23,100
4	T-shirts etc	'000 pieces	8,105	6,614	32,800
5	Jerseys etc	'000 pieces	7,610	6,541	8,250
6	Shorts, trousers	'000 pieces	11,370	8,297	9,650
7	Blouses	'000 pieces	2,629	1,946	2,850
8	Shirts	'000 pieces	11,397	7,785	9,000
10	Gloves, mittens	'000 pairs	37,001	..	42,500
21	Parkas, anoraks	'000 pieces	6,275	5,173	6,500
26	Dresses	'000 pieces	2,461	2,766	3,200

Source: *Textile Asia* (January 1989), p. 14.

Although China obtained a considerable increase in quotas on textile items they are small relative to the country's capacity to export textile products. Hence, China has been pressing for substantial improvement in its access to the EC markets. As against this, the new Four-Year MFA between China and the EC, which came into force in 1989, introduced new restrictions on synthetic yarn and knitted or crocheted gloves as well as for suits and socks of man-made fibres. China is the largest MFA supplier of textiles and clothing to the EC in volume terms. Since quota restrictions are imposed on quantity, higher prices offering "scarcity rent" could come only from high quality products. While China could continue to reap the benefits of relatively low prices of its non-quota items, restricted items will need to be specifically identified as target product areas for substantial improvements in quality in view of the large gap that persists between aspirations and actual implementation in liberalizing the global textile and clothing trade.

C. IRON AND STEEL: REVEALING THE SUPPLY GAP

The resource base

China abounds in iron ore with 50 billion tons of proven reserves. The geographic distribution of iron ore mines is somewhat skewed with two-thirds of reserves lying in northern China and the remainder in the south. The Anben area in north-east China has the most productive deposits, and its reserves account for one-fourth of China's total iron ore reserves. However, deposits are scattered in 27 provinces, autonomous regions and municipalities.

The grade of iron ore deposits is low and most of them have to be beneficiated. According to rough estimates, about 98 per cent of iron ore reserves are of low quality with an average iron content of around 30 per cent. China is endowed with rich resources for refractories, e.g., high grade bauxite, magnesite, flint clay. An added advantage stems from around 954 billion tons of proven coal reserves, with one quarter being coking coal. Other pertinent resources include nickel, molybdenum, and manganese for refractories. It is believed that with an advanced ore dressing technique, it should be possible to turn out concentrates with an average iron content of around 65 per cent.

The growth of iron ore mining has failed to keep pace with the needs of the country's steel industry. Huge quantities of iron ore imports are inevitable to ease the supply shortage. Iron ore output increased by 58 per cent in 1981-1988, compared with a 66 per cent rise in steel output during the same period. Iron ore production currently stands at around 160 million tpy, and estimated to rise to 185 million tpy in 1995 and 200 million tpy in the year 2000. It will be insufficient to meet the envisaged steel output of 90 million tons by the year 2000. An additional annual output of over 50 million tons will be required to cope with the rising steel production in the 1990s. Imports of high-grade iron ore, currently at around 10 million tpy, are likely to increase to 20 million tpy in the near future.

Emerging trends

China's iron and steel production rose significantly from 20.46 million tons in 1976 to 59.43 million tons in 1988. Since 1982, the annual output has increased by over 3 million tons leading to a record high output of 61.32 million tons in 1989 despite shortage of coal, electricity, transport facilities, funds and other constraints. China ranks fourth in the world production of iron and steel after the Soviet Union, Japan and the United States. However, domestic supply is short of demand. In 1989 China imported 9.13 million tons of steel at a cost of $4.43 billion.

Fig.IV.3. Production of pig iron, steel and rolled steel, 1980 and 1990 (million tons)

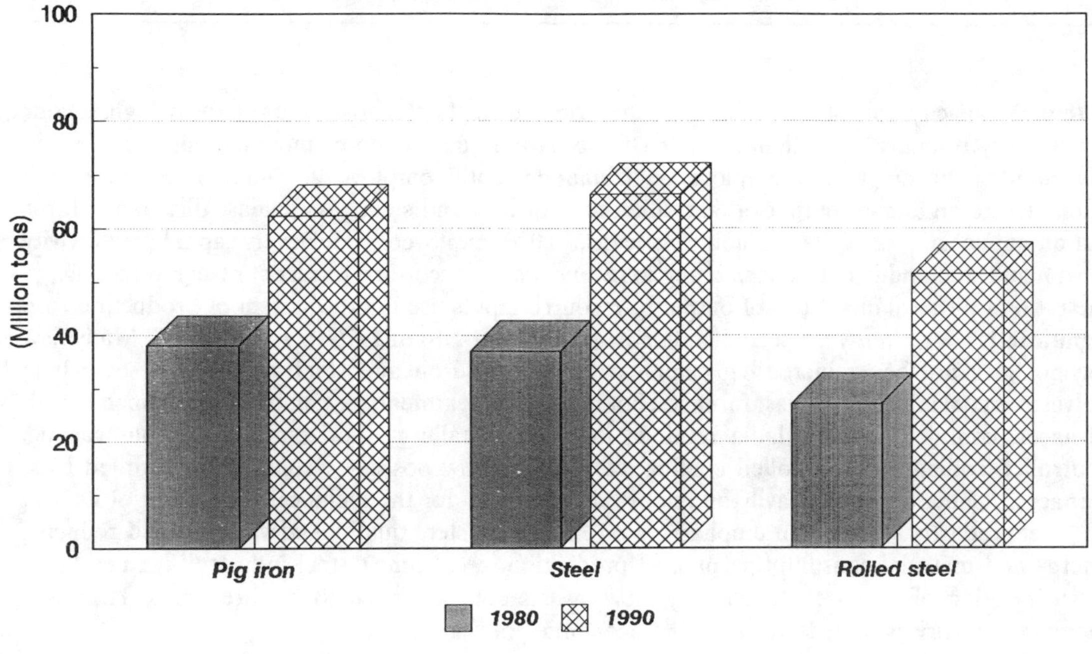

In 1990 there were around 1,400 iron and steel enterprises, including 15 large complexes each with an annual capacity of 1 million tons, such as the Wuhan, Anshan, Baotou, Panzhihua, Baoshan and Shoudu iron and steel companies. Except for Tibet, the iron and steel industry is spread nationwide. With a diversified structure, China is capable of producing over 1,000 types of steel products, and around 20,000 types of rolled steel.[9/] Products manufactured by the State-owned smaller steelworks are presented in Annex Table A-16. In 1987 these small enterprises produced 22.25 million tons of iron and steel.

Almost all product varieties are in short supply. For example, domestic production capacity of steel sheet is 8 million tpy against an annual demand for 10 million tpy, estimated to reach 14-15 million tpy in 1995.[10/] The production capacity of cold-rolled steel sheet is 2.5 million tpy, meeting just half the annual demand. The production capacity of 2.5-4.0 million hot-rolled steel sheet is also limited, depending heavily on imports. With the price tending to fall in the face of a continued slump in processing activities in the early 1990s,[11/] the domestic production of steel might suffer serious losses due to the soaring prices of raw materials and energy.

Industrial slow-down in 1990 did not strike a balance between demand and supply, and steel products needed for basic industries such as energy, transportation and raw materials continued to be in acute shortage. Table IV.10 reveals the serious shortage of selected steel products and the extent of the demand which they represent.

Table IV.10. Shortage of selected steel products, 1990

Product	Shortage (tons)	Percentage of demand
Light track	45,100	56.3
Wire rod	242,400	71.4
Medium plate	414,400	72.3
Sheet steel	230,000	83.0

Source: Economic Information Department, *Trade Promotion* (June 16,1990), p. 6.

Although capacity utilization rates across the iron and steel enterprises are relatively high, labour productivity remains lower than in their counterparts in developed countries. This is evident in the number of employees employed. China currently employs 2.5 million employees to manufacture an annual output of 60 million tons of iron and steel. As against this, Japan turns out an annual output of 100 million tons with 250,000 employees. However, capital productivity in iron and steel industry has been rising in recent years. According to recent research findings[12/] the ratio of physical quantities of output to resource inputs used in the system of production rose significantly over a ten-year period ending in 1985 - from 1.10 in 1976 to 1.43 in 1985. Many new technologies are being increasingly applied in Chinese iron and steel works. These include pulverized coal injection in blast furnaces, hot metal pre-treatment, oxygen blowing on open hearth furnaces, complex blowing, ladle refining, injection metallurgy, horizontal continuous casting, controlled rolling and controlled cooling, etc. China now possesses computer-controlled blast furnaces. China depends heavily on imported technology for the technical renovation of its iron and steel industry. The main emphasis is on improving steel quality and varieties and reducing energy and material consumption in steel production. As China's steel industry is in need of a fresh injection of advanced technology, new investments will need to be directed to large and medium enterprises in order to reap the economies of scale.

Baosteel complex: a model for technical renovation

The Baoshan General Iron and Steel Works (Baosteel) represents China's largest iron and steel project undertaken since the founding of the country. This fully integrated plant was commissioned in 1985 with an annual target capacity of 3.18 million tons in Phase I. With the completion of Phase II, total annual capacity of Baosteel will ultimately be increased to 6.7 million tons.

Baosteel is located north of Shanghai on the banks of the Yangtze river. The complex is based primarily on technology supplied by Japan and the Federal Republic of Germany. Most of the raw materials and fuel are transported to Baosteel by water. At full capacity utilization around 20 million tons of material will have to be handled annually. Coal is sourced from domestic mines, while the major sources of iron ore are Australia, Brazil and India.

Phase I of Baosteel was built at a cost of $4.18 billion, with $2.78 billion on imported equipment. The total investment in the second stage is expected to be doubled. Most of the output of Phase I was semi-finished crude steel and only 20 per cent was finished steel pipe. On completion of Phase II in 1990 the proportion of crude steel was expected to be reduced to around 20 per cent and about 80 per cent will be steel plate, pipe, and other steel products. The plant is operating normally, with output rising steadily and quality reportedly up to the expected standard.

The technological sophistication of Baosteel is well advanced to serve as a model for other steel complexes and large enterprises. The plant now encompasses blast furnaces, oxygen top-blown converters of 300 tons capacity provided with a hot metal desulphurization station, and the desulphurization unit capacity of the old enterprises through technological renovation.

Prospects

An upward trend in oil prices came at a time when the world steel industry was recovering from the problem of overcapacity. Ever since the first oil shock, energy conservation had been a priority in steel production. In China oil is rarely used as an alternative to coke in blast furnaces. On the demand side there has been a continuous upward surge in China. In order to strike a balance between supply and demand, China could endeavour to reduce the steel intensity of industrial production and look for alternative means. This could help China save substantial foreign exchange currently being spent on steel imports.

In rehabilitating existing mills priority is accorded to upgrading of large steel mills with an annual production capacity of 1 million tons, special steel plants and plants producing mining equipment. In the production process, emphasis is placed on improvements in steel-making technology and major rolling machines. The manufacture of complete sets of equipment and support expansion projects are also planned for technical transformation. The China International Iron and Steel Investment Corporation undertakes the task of facilitating the flow of foreign funds into plants and projects in the iron and steel industry.

It is proving to be cheaper and quicker to buy foreign idle steel-making equipment and erect it within China. Hence China is considering buying redundant foreign steel works of advanced design. The purchase list presented by China includes hot- or cold-rolling mills for steel and rolling mills for pipe and wire rod. The tendency is to buy foreign plants mainly from Japan where steel-making facilities often turn out to be idle because of constant restructuring plans. Steel-making technology is well advanced in some developing countries. China could negotiate with those developing countries for mutual benefit. The total steel output is expected to reach 72 million tons in 1995 and around 90 million tons in 2000. By the year 2000, China will endeavour to introduce the world's most advanced steel-making technology and to build one or two of the most modern steel production bases.

D. CHEMICALS: EXPANDING FROM INFANCY

The resource base

China is endowed with an abundant raw material resource base for the manufacture of acid and soda, chemical fertilizers such as nitrogenous compounds, phosphate and potash, and a wide range of organic chemicals, especially petrochemicals. Troilite deposits, used in the production of sulphuric acid, are located mainly in Gansu and Guangdong provinces, followed by Hunan, Shanxi and Anhui provinces, which are known for their well-developed non-ferrous metal and chemical industries. Salt being the main raw material for the production of sodium carbonate, the production units of sodium carbonate are located near salt producing areas which are endowed with reserves of limestone and an adequate supply of electricity. Coastal areas such as Dalian, Tianjin and Qingdao are the main sodium carbonate producing areas. Sodium carbonate and limestone constitute the main raw materials for the manufacture of caustic soda. Limestone reserves are widely distributed in China.

The main raw material for the manufacture of nitrogenous fertilizer is synthetic ammonia which is produced mainly from coal, petroleum and natural gas. Coal is China's premier energy resource. The country has the world's second largest proven coal reserves of 954 billion tons, and ranks very closely behind the Soviet Union. With an output of 1.05 billion tons of raw coal in 1989, China became the largest coal producer in the world. Between 1949 and 1988 China opened and expanded 2,130 coal mines. The target of 1.4 billion tons of raw coal is set for the year 2000. Proven reserves of oil are large, but natural gas is rather scarce in China representing only 2.1 per cent of Chinese energy output.[13/]

One of the main raw materials required for the manufacture of phosphate fertilizers is phosphate. China encompasses over 3 billion tons of phosphate rock with a high magnesium oxide content. The government endeavours to exploit the major deposits of phosphate rock reserves particularly in Guizhou, Hubei, Yunnan and Sichuan provinces. Transport bottle-necks constitute the major constraint on phosphate production.

The principal organic chemical, calcium carbide, is manufactured in China from coal and coke and is turned into synthetic organic products. Oil and gas also are important raw materials for varieties of organic chemicals. Plastics, synthetic rubber and synthetic fibre production units are located mostly near oilfields and oil refineries.

Emerging trends

The gross output value of the chemical industry grew at an average annual rate of 14 per cent during 1953-1986, compared with the industrial average growth rate of 11 per cent.[14/] Despite rapid output growth in recent years, China remains the world's largest chemical importer and the chemical industry is just expanding from infancy.

Chemical plants were first constructed in China during the early 1940s and usually incorporated Japanese technology. During the 1950s they were redesigned by Soviet technicians and after their withdrawal technical improvisation slowed down. After the inauguration of the "Open Door" policy in 1978, joint ventures with western transnational corporations (TNCs) were largely sought to speed up modernization of chemical plants. Sinochem, the country's largest and oldest company, played a crucial role in importing chemical plant technology during the 1970s and 1980s.

During the Seventh Plan (1986-1990) priority was accorded to three areas within the chemical subsectors: (a) rapid growth of compound fertilizer production, (b) substantial expansion of production capacity for soda ash and caustic soda, and (c) expansion of production of fine chemicals such as pesticides, dyestuffs, adhesives and catalysts. The Seventh Plan placed emphasis on a rapid reorganization of industry and an extension of enterprises autonomy within the chemical branch. The introduction of the contract responsibility system has facilitated decentralization of authority to lower levels of management. Nevertheless, the government retains firm control over the development of the industry.

The production of major chemical products stage a significant recovery during 1986-1990 after suffering a fall in output of some products in 1984 and 1985 (see Annex Table A-17). Having suffered a drastic fall in output for two consecutive years, the production of sulphuric acid rose significantly during 1986-1990. The demand for sulphuric acid increased sharply because of the rapid increase in phosphate fertilizer production. However, some small units could not come into operation immediately because of obsolete equipment, and the majority of manufacturing plants based on sulphur could not operate at full capacity utilization level when the government abolished subsidies on sulphur imports.

There has been a constant and steady increase in the output of soda ash and caustic soda over the years. The growth rate of small-scale soda ash enterprises surpassed that of the major large-scale enterprises in recent years. The increase in output is attributed mainly to improvements in the production level of existing plants. Gong Prefecture Soda Plant in Henan, Juigchag Chemical Complex in Gansu and Zhangjiagang Municipal Chemical Fertilizer Plant were completed and came on stream by end-1985. The quality of the product from large-scale enterprises was also significantly increased.

The output of chemical fertilizers (nitrogenous and phosphate) recovered markedly from severe declines in 1985. By 1990 the production of chemical fertilizers was 187.9 million tons compared with 12.4 million tons in 1981 (see Annex Table A-17). The decline in the production of chemical fertilizers in 1985 was due to the government's decision to close around 200 small inefficient plants producing low quality products. China continued to expand the production capacity for high-grade fertilizers, particularly phosphate fertilizer. Around 17 per cent of apparent consumption of chemical fertilizers was met by imports in the mid-1980s.

Annex Table A-17 shows a drastic fall in the production of chemical pesticides over the years. The 1990 output was 229,000 tons - less than half of the 1981 production level. The present demand for high-quality pesticides is 25-50 per cent higher than the available supply. Agriculture in China was adversely affected by insects in 1988. Some 100 million paulownia trees in the plains of Henan have been partially defoliated by bag worms. Around 30 million hectares representing 30 per cent of the country's arable land are treated with pesticides. China thus represents one of the largest pesticide markets in the world.

The production of ethylene rose rapidly in recent years. China has completed building four ethylene plants with a combined capacity of 1.2 million tons a year. In 1990 China approved an ethylene project to be set up by the Taiwan Plastic Corporation (TPC). With added ethylene capacity, production is expected rise significantly to 2.26 million tons in the near future. Although China's petrochemical industry dates back to the 1960s many organic-chemical technologies have emerged. Many of the new devices were practised in small- and medium-sized plants in the 1970s. In the 1980s many petrochemical complexes were constructed with foreign technology, but a few units are still based on domestic know-how. In synthetic resins and plastics, acetylene chlorination is the main means for making polyvinyl chloride (PVC).

Investment trends

Investment in fixed assets has grown at an annual average rate of about 20 per cent during 1985-1989; the government's share of this investment being roughly 45 per cent. It is clear in relative terms that chemicals do not represent a high priority investment area, and their share in total industrial capital construction expenditure has been about 7 per cent in recent years. The government's share of modernization investment within the chemical sector has usually been below 5 per cent. Industry employs about 300,000 workers annually and employment has grown very slowly at the rate of 2.8 per cent per annum during 1978-1988. This testifies to the increasing capital intensity of the chemical industry. Chemical production remains concentrated in the coastal provinces, which accounted for around 60 per cent of the gross value of the industry's output in 1986 - there is virtually no chemical production in the western regions.

Four giant plants with assets in excess of Rmb 1 billion account for over 20 per cent of the gross value of industrial output. Sixty other plants with assets in excess of Rmb 300 million account for another 30 per cent. As against this, 300 plants with assets ranging between Rmb 30 to Rmb 50 million produced only 28 per cent of aggregate branch gross output value in 1987. Chinese Petrochemical Corporation (Sinopec), formed in 1983, is evidently a rising force in the country's chemical industry. The Corporation runs huge new petrochemical plants which are often referred to by foreign observers as the "show cases" of the industry. Complexes such as Yanshan, Liaohua and the Daqing Ethylene plant are the product of foreign expertise, but like the new petrochemical complexes in Saudi Arabia, they show every sign of being neatly and efficiently run by domestic technicians. The plants are much more advanced than those of the Chinese industry as a whole. In 1988, Sinopec had 24 plants on stream and had test runs at 37 plants manufacturing various petrochemicals. In the same year Sinopec processed over 90 per cent of China's crude oil. The Corporation produced 1.11 million tons of ethylene, 1.0 million tons of synthetic resins and plastics and 190,000 tons of synthetic rubber. In 1988, Sinopec's Qilu Petrochemical Co. completed China's largest polyvinyl chloride plant with a capacity of 200,000 tons a year. Trial production of PVC stood at 30,000 tons.

Joint ventures are very few in the Chinese chemical industry. A pharmaceuticals joint venture at Shanghai with the United States company Squibb and a trading joint venture with the German trader Karl O. Helm, based not in China but in Hamburg and geared to promote Chinese exports to Europe, are the main ones. While joint ventures generally look at the domestic market potential, many local firms have recorded significant export successes. For instance, the Warrior

Fig.IV.4. Physical output of selected chemicals, 1981 and 1990 (Million tons)

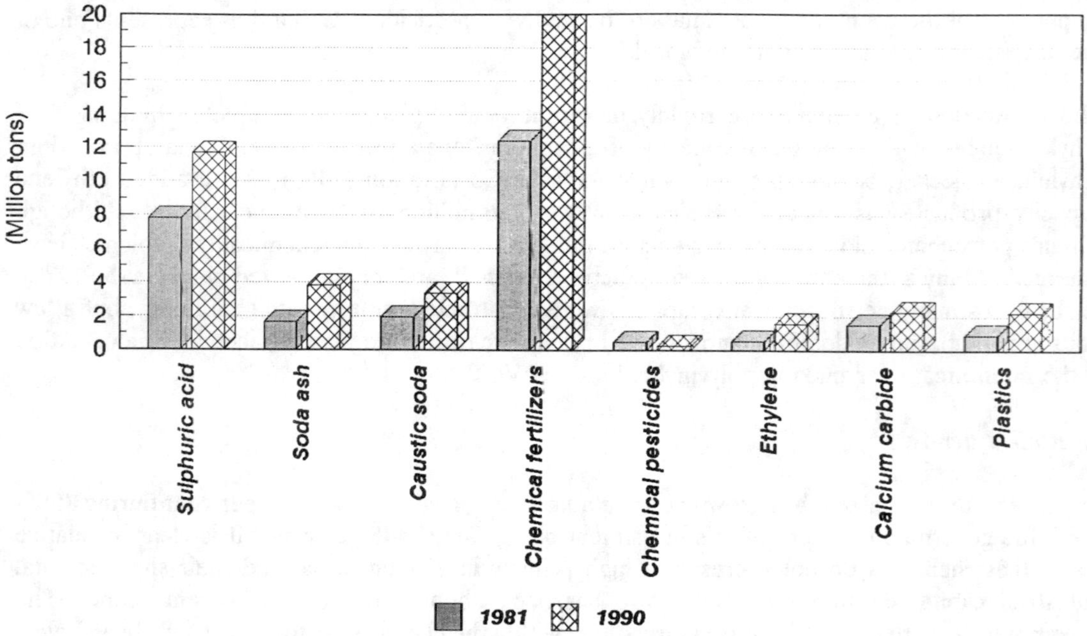

factory, located in the heart of Shanghai, has won the United States Department of Trade and EC certification as well as local gold awards for quality. Joint ventures with foreign partners have a very high priority in official plans for China's chemical industry as the industry badly needs further infusions of foreign know-how and technology. The partner is expected to import the plant and equipment for the venture, easing the burden on the scarce foreign exchange revenues. This means that anything imported by the venture, whether raw materials or equipment, must be paid for by foreign exchange generated by the venture itself. Thus the venture has to be export-oriented. This also implies that to the extent that foreign firms are exporting their production to earn foreign exchange, they are allowed to use China as a manufacturing base. But most multinationals do not seem to look at China as a manufacturing base for accelerating exports, but look at China as an unlimited market for chemicals.

Selected chemicals

Fertilizers

Despite a significant increase in production capacity, China continues to import chemical fertilizers due to continued supply shortages. China imported $2 billion worth of fertilizers during the first ten months of 1989, representing a 14 per cent increase over the corresponding period in 1988.[15/] According to official estimates, domestic production of fertilizers in the first 10 months of 1989 was around 15 per cent short of target. Imports of fertilizers rose from 13.9 million tons in 1989 to 16.3 million tons in 1990. An additional 1.8 million tons of fertilizer were expected to be used in the central, north-west and south-west regions in 1990.

Table IV.11. Current large-size fertilizer projects as of May 1990

Serial No.	Project	Province	Product	'000 tpy	Nitrogenous '000 tpy	Feed-stock	Start-up	Source of finance
1	Hejiang	Sichuan	NH3/urea	300/520	240	NG	1992	Italy
2	Fuling	Sichuan	NH3/urea	300/520	240	NG	1992	France
3	Jinxi	Liaoning	NH3/urea	300/520	240	NG	1992	France
4	Dalian	Liaoning	NH3	300	(250)	Oil	1992	Germany, Federal Republic of
5	Weihe	Shanxi	NH3/urea	300/520	240	Coal	1993	OECF
6	Jiujiang	Jiangxi	NH3/urea	300/520	240	Oil	1993	OECF
7	Jilin	Jilin	NH3/urea	300/520	240	Oil	1993	ADB
8	Neimenggu	Neimenggu	NH3/urea	300/520	240	Oil	1992	OECF
9	Chengdu	Sichuan	NH3	200	160	NG	1991	Local
Total					1840			
10	Dayukou	Hubei	TSP	560	270		1993	IBRD
11	Huangmailing	Hubei	MAP	140	93		1993	IBRD
12	Wengfu	Guizhou	TSP	800	380		1995	OECF
13	Anning	Yunnan	TSP	400	190		1995	OECF
14	Yunfu	Guangdong	TSP	400	190		1994	IBRD
15	Luzhai	Guanxi	DAP	240	110		1995	OECF
16	Qinhuangdao	Hebei	DAP	480	220		1992	JV Kuwait
17	Jining	Shandong	DAP	480	220		1995	ADB
Total					1673			
18	Chaerhan	Qinghai	KC1	800	500		1996	IBRD

Source: Bruce Stone, *Fertilizer Use Development in China During the Reform Period*, paper prepared for the World Bank, FAO, UNIDO Industry Working Group, May 21-23, 1990.

The government's main strategy is to achieve a desirable balance between fertilizer supply and demand by rapidly expanding the production capacity of phosphate fertilizers. the main objective is to move towards self-sufficiency by the year 2000. Table IV.11 furnishes details pertaining to large0size fertilizer projects as of May 1990. All these projects are likely to come on stream in the 1990s, adding over 10 million tons of additional capacity to China's fertilzier industry. Of the 18 projects, 17 projects are financed by bilateral and multilateral sources, while only one project is locally financed.

Optimal capacity utilization in these large-scale fertilizer projects may significantly ease supply shortage in the 1990s. Capacity utilization is constrained by infrastructural bottle-necks and energy shortage. The demand for fertilizers continues to rise as farmers try to enhance the yield per hectare. Given the limited availability of arable land, further increase in agricultural production could be enhanced only through intensive cultivation using heavy dose of fertilizers. In the application of fertilizer inputs in China, there is already some imbalance with more than proportionate use of nitrogenous fertilizer (see Table IV.12). In the search for greater agricultural productivity the current emphasis is on the use of other fertilizer nutrients.

Table IV.12. Chemical fertilizers applied in agriculture by region, 1988-1990
(Thousand tons)

Region	Nitrogenous			Phosphate			Potash			Complex		
	1988	1989	1990	1988	1989	1990	1988	1989	1990	1988	1989	1990
Beijing	..	86	98	10	9	9	..	..	1	15	22	36
Tianjin	43	47	47	4	6	6	1	2	3	9	12	13
Hebei	833	885	921	242	249	261	18	22	26	163	200	244
Shanxi	297	310	344	87	103	131	7	7	10	53	68	81
Inner Mongolia	164	188	209	43	48	54	4	7	7	49	49	75
Liaoning	500	537	574	121	119	132	5	8	11	75	82	97
Jilin	422	462	548	35	33	36	13	16	24	197	204	239
Heilongjiang	265	308	370	86	157	171	14	18	24	179	178	200
Shanghai	134	148	161	29	27	80	1	1	2	3	4	5
Jiangsu	1381	1458	1506	332	340	362	32	36	68	220	213	282
Zhejiang	706	697	678	123	125	138	37	44	54	55	59	77
Anhui	811	886	904	289	304	319	39	40	51	127	142	171
Fujian	416	440	437	113	135	128	83	104	120	59	70	79
Jiangxi	429	435	461	160	161	178	104	113	133	45	52	64
Shandong	1373	1479	1541	309	329	354	41	50	70	349	417	490
Henan	1012	1234	1373	363	433	520	30	51	60	101	125	179
Hubei	842	843	912	245	274	311	53	66	73	100	134	190
Hunan	794	770	770	223	229	213	139	161	176	70	84	102
Guangdong	882	944	958	183	219	201	187	224	276	130	176	189
Guangxi	373	400	450	128	131	148	114	129	163	63	75	101
Hainan	68	73	70	13	17	19	10	11	9	16	19	24
Sichuan	1070	1263	1391	315	316	371	25	20	29	88	101	134
Guizhou	230	244	263	70	66	72	10	20	22	20	25	30
Yunnan	285	325	355	110	120	115	10	22	28	38	54	57
Tibet	6	5	7	1	2	4	..	1	..	5	5	5
Shaanxi	366	425	486	67	83	101	13	17	23	49	59	69
Gansu	144	185	216	48	70	92	7	7	5	44	62	62
Qinghai	20	24	27	8	9	12	2	3	2	14	15	12
Ningxia	60	74	82	12	14	13	..	..	..	21	23	21
Xinjiang	164	193	225	52	61	73	5	5	9	55	80	88

Source: State Statistical Bureau of China, *China Statistical Yearbook* (various issues).

Under normal conditions the fertilizer stock, stored mainly during the low season, amounts to 25 per cent of total output sold. Most of the fertilizer distributed from the central level is subsidized and sold at a uniform retail price throughout the country. There has been no price change since 1984. Farmers purchase fertilizers with the aid of purchase certificates (coupons) issued by the Adminstration in exchange for farm products sold to the State. According to rough estimates, for every 100 kg of grain sold to the State, the farmers are entitled 1.5 kg of urea purchase certificates. The government has recently indicated that farmers must rely less on support for input purchases.

Petrochemicals

The newly emerging petrochemical industry of China has achieved significant strides in capacity expansion and production. Production data for selected petrochemicals depict the vibrant growth record of the industry in the second half of the 1980s (see Annex Table A-18). Most of the enterprises are under the China Petrochemical Corporation. The Corporation has forged ahead with the reform of technical systems and extensively integrated scientific research with production. However, many petrochemical plants are in need of technical renovation.

China's ethylene industry had a production capacity of around 1.6 million tons in 1988. Actual production stood at 1.1 million tons in the same year, leaving around half a million tons of excess capacity. Idle capacity in the ethylene industry is due largely to inadequate supply of oil, the chief raw material. According to rough estimates around 6.5 million tons of crude oil will be needed to increase ethylene production. Such a volume of crude oil represents around 5 per cent of China's estimated oil flow. In the wake of acute energy shortage, it is unlikely that such a proportion of a daily output of about 60,000 barrels would be destined for ethylene production in the 1990s. The fastest way to boost ethylene output is to undertake technical renovation on the capital stock in order to achieve productivity gains.

The major petroleum-routed synthetic resins produced by the China Petrochemical Corporation are polyethylene, polypropylene and polystyrene. Total synthetic resins production in 1988 was 541,000 tons more than that of 1984. In the mid-1980s, the enterprises under the China Petrochemical Corporation significantly reduced production costs largely through energy conservation devices, and raised the quality of products to comparable international standards.

Synthetic rubber production rose significantly from 128,000 tons in 1984 to 184,000 tons in 1988. In 1986, there were 6 synthetic rubber factories under the China Petrochemical Corporation with a total annual capacity of 159,000, representing 63 per cent of the overall national synthetic rubber capacity. The consumption of synthetic rubber in China is almost equal to the production of natural rubber, implying that the barriers between the two categories of rubber are being increasingly dismantled, and the complementarity as opposed to the substitutional aspect enhanced. The fact that major natural rubber producers in Asia experienced the fastest growth in the consumption of synthetic rubber in the 1980s, lends credence to the increasing inroads of synthetic rubber in the production of rubber products.

Significant strides have been achieved by the China Petrochemical Corporation in the production of synthetic fibres. In the mid-1980s, the Corporation had 16 synthetic fibre units under its control. A number of them were fibre complexes such as the Shanghai General Petrochemical Works, the Liaoyang Petrochemical Fibre Company and the Tianjin Petrochemical Company. The quality of the high concentration acrylic fibres developed by the Shanghai complex has proved to be close to world standards.

Constraints and prospects

Despite gluts in the world market because of weakening demand and overcapacity, the demand for and production capacity of basic chemicals are increasing in China. However, such expanding scales are not without their blemishes.

Shanghai highlights the problems facing the older chemical complexes in cities such as Beijing and Nanjing. Shanghai is still a major centre of China's chemical industry, producing 29 per cent of the country's chemical fibres and 15 per cent of its plastics in the late 1980s. But it has stagnated in recent years, compared to neighbouring Jiangsu province and other new economic zones. One of the most decisive factors weakening Shanghai in comparison to other cities and provinces is that about half of the city's industrial equipment pre-dates 1950, with a further 33 per cent installed before 1960. Shanghai has spent more than $2 billion a year in recent years to renovate industries and build new manufacturing facilities. But the task of large-scale renovation of existing, well-established industry is in most cases more difficult than building new industry.

Older Chinese factories contain a mix of technologies acquired from different countries. Immediately after the Second World War, the country copied details of German chemical engineering technology published in the famous Bios Report. During the 1950s, the Chinese relied on the Russians. In the 1970s, the Chinese bought equipment and licensed processes mainly from the Federal Republic of Germany, Japan and the United States, and since the opening up China has made the most of the freedom to shop around for the best quality and price. One of China's newest factories built at a cost of Rmb 270 million for manufacturing fine chemicals for paper, leather, and textile coatings, acrylic paints and resins, was stocked with Mitsubishi computers. A Japanese catalyst company provided the process, but the know-how for the finishing products came from Union Carbide in the United States.

Infrastructural bottle-necks hamper the rapid expansion of the chemical industry in China. Although China is well endowed with a good raw material resource base for the chemical industry, inadequate transport facilities have restrained the optimal use of the resource base. The prospects for chemical products are limited to energy supplies. Even during the current phase of slow industrial growth in 1990, the Chinese economy faced the threat of energy shortages. The problem will need to be alleviated by energy conservation measures by users. But this needs substantial investment in technical renovation.

While the world chemical industry is faced with the new wave of oil shocks raising doubts as to how much of the extra costs can be passed on to consumers given the glut, domestically the prospects remain good for the chemical industry in the face of increasing demand in the huge Chinese market.

E. BUILDING MATERIALS: FOREIGN TECHNOLOGY FOR PRODUCTIVITY GAINS

An overview

China has a diverse mineral resource trove for the production of building materials. Major products of the building materials industry in China are cement and cement products, sheet glass, sanitary ware, bricks, tiles, gypsum board, aeroconcrete, hermetically sealed waterproof materials, furnishing material, asbestos, graphite, gypsum, talcum, porcelain clay, mica, diamonds, and other non-metallic minerals and mineral products. Other building materials include organic non-metallic minerals such as special glass, special porcelain, quartz glass complex materials, synthetic crystals, ingot stone, special inorganic glue materials, etc. There are over 220,000 enterprises manufacturing these building materials. During 1985-1988, the building materials industry employed on average around 0.3 million workers per annum.

The major building material plants are mainly located in the provinces of Jiangsu, Shandong, Zheizang, Guangdong and Hubei - together these provinces account for only about 25 per cent of the total value of industrial output, however. This reflects the relatively wide regional dispersion of the building materials industry.

There had been a continued steady growth of China's building materials industry during 1979-1987. With a significant annual increase in the output of cement, China's cement production ranked first in the world for three consecutive years. By 1987, the output of cement stood at 180 million tons, sheet glass at 56.01 million standard cases, sanitary ware at 11.86 million pieces, and glazed well and ground bricks at 76.9 million square metres. Net industrial output value (a concept close to though not identifiable with value added) represented about 35 per cent of gross output value during 1985-1987. The value of this ratio was generally lower for most other industrial branches. Net industrial output value[16/] has, however, grown at an annual average of about 7 per cent during 1985-1988 which is significantly below the rate of growth of gross industrial value.

Since 1988, the expansion of the building materials industry has been seriously affected by the freezing of major construction projects in the larger Chinese towns. A total of 37 construction projects were halted in Beijing. The construction of seven joint venture hotels was halted in Canton, where the market is flooded. But foreign investment projects have, generally been hit less hard by the austerity programmes than locally financed ones so far, largely because China is trying to establish an image as a safe place for investment.

In order to offset the decline in domestic demand, China is seeking to expand cement exports - exports are still however well short of 1 per cent of production. The Chinese cement industry expects cement demand to exceed 250 million tons per year by the year 2000. To meet this demand, China has been seriously investigating western technology and equipment. The country's cement industry also intends to broaden the use of computers, further explore the use of mineralizers in the burning of clinker, utilize industrial wastes on a wider scale, and speed up tests on fluidized bed furnaces for the burning of cement clinker on an industrial scale.

Fig.IV.5. Volume of cement production, 1985 - 1990 (Million tons)

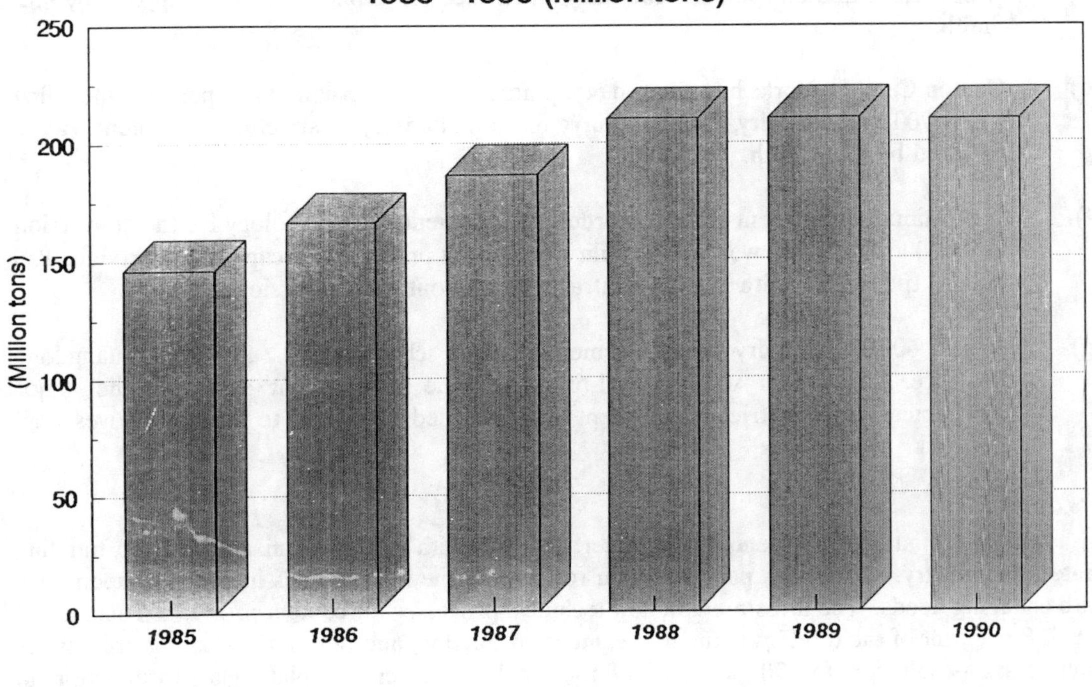

Recently, China completed and put on-line at least one modern cement plant. This was the new, 3,200 tons per day (tpd) plant for Guangxi Building Materials Export Supply Co., at Liuzhou. The plant went on stream in 1988. Major equipment for the plant was supplied by F.L. Smith, including a dry process kiln with pre-calciner and Folax cooler, two Combidan cement mills, a Tirax coal mill, and a Duodan raw mill.

Additionally, the following plants came on-line in 1988 and 1989. The Jin Shan plant near Shanghai completed a conversion from wet process to a four-stage SP kiln. The conversion will almost double the capacity of the previous line. F.L. Smith is handling the work. The 600,000 million tpy plant in Yaoxian near Xian Shanxi was completed in 1987. The plant has a DD furnace and is scheduled to receive a training simulator. Japan's Kobe Steel is supplying both the furnace and the simulator. The Guangzhou Municipal Bureau of Building Materials plant near Canton is a new 4,000 tpd plant with equipment supplied by F.L. Smith, including a five-stage SLC pre-calciner kiln with Folax cooler, a Duodan raw mill, Tirax coal mill, and two Combidan cement mills. The plant came on-line in the first half of 1988.

A new 2,000 tpd plant is planned by the Fujian Provincial Government at Shunchang. The plant will be supplied by F.L. Smith, and the equipment includes a pre-calciner kiln with Folax cooler, an Atox raw mill, two Combidan cement mills, and a Tirax coal mill. Commissioning began for 1988. Several projects in various planning or development stages during 1989 included:

(a) A new 4,000 tpd plant for Tongling (Anhui Province), and a 2,000 tpd, dry process line extension for the Zhong Guo plant. Germany's Dyckerhoff Engineering has already done the initial field work for the feasibility studies on both of these projects.

(b) The Yan Shan plant near Beijing continues work on its project to expand the plant by 700 million tpd. F.L. Smith is supplying the major equipment.

(c) A 2,000 tpd plant still planned for Lunan in Shandong Province, but no further details have been announced on the project. F.L. Smith will supply the equipment.

(d) A new 2,000 million tpd plant for Jiangxi Province. Equipment is to be supplied by F.L. Smith.

(e) Canton Cement Works has ordered equipment for an expansion of its operation including a new 700 tpd, semi-dry, ILC kiln slurry filtration and drying system. Equipment will be supplied by F.L. Smith.

(f) The Baimashan Cement plant has ordered equipment and technology for the conversion of its 600 million tpd wet process kiln to semi-dry process with capacity to produce 850 million tpd. F.L. Smith has the contract to carry out the conversion.

(g) A new 600,000 tpy, dry process cement plant is scheduled at Yun Fu in Guangdong Province. The plant will be coal-fired, and the contract for supplying the major mechanical and electrical equipment was awarded last year to France's Fives Cail Babcock.

In addition, China has in recent years concerned itself with the modernization of the building materials industry. China has paid attention to adopting new techniques in order to keep pace with changing needs. The investment in new technical processes for cement production accounted for 72.8 per cent of the total investment in cement production, and that in the floating process for plate glass production for 70 per cent of the total investment of plate glass production in 1986-1987. A production line using modern technical processes with an annual capacity of one million tons of cement went into operation at the Liuzhou Cement Plant in Guangxi. A new production line using floating process with an annual capacity of 2.3 million cases began production at a Guangdong Float Glass Works. Another production line for porcelain sanitary ware also went into production in Guangzhou. Three production areas for complete porcelain toilet sets were built in Tangshan, Shanghai and Guangzhou. Complete sets for three star hotels are no longer imported. By the end of 1987, technical transformation began in all old cement plants built before 1949 as well as part of the old cement plants built in the 1950s.

Performance

The average grade of cement has improved markedly, but the average grade of clinker has only improved slowly during the period 1979-1987. The rate of first grade glass plate in total production has actually declined from 82 per cent in 1985 to 69 per cent in 1987. Most of the energy requirements of the cement industry are still supplied by coal. But electricity is gradually being substituted for coal. Whereas the kg/ton ratio of coal consumption within the cement industry has declined from 201 in 1985 to 193.3 in 1987, the kWh/ton consumption of electricity has increased from 103.9 to 106.4 over the same period. The utilization of industrial waste for manufacturing cement is low. Plate glass production is becoming more energy intensive over time. Coal consumption for plate glass production has increased from 30.76 kg per weight tare in 1985 to 32.66 kg per weight tare in 1987. Electricity consumption has increased from 5.26 kWh per weight tare to 5.37 kWh per weight tare over the same period.

The slow pace of improvement in product quality is also reflected in the equally slow growth of factor productivity within the industry. During 1984-1987, labour productivity remained stagnant in the cement branch - 254 tons of cement was produced per employee annually throughout this period. The operating ratio of rotary kilns declined from 84.07 per cent in 1984 to 80.77 per cent in 1987. The rate of surplus (profits to taxes) to the value of fixed assets averaged a little over 15 per cent for the 1985-1987 period for the building materials industry - significantly below the value of this ratio for industry as a whole as well as for heavy industry. This is despite the fact that the cement industry has benefited from a reduction in taxes and the "de-freezing" of a disproportionately large number of prices. Many input prices are however also now determined by market forces. This has created pressure within the industry, and throughout the 1980s there has been a growing number of loss-making enterprises. These were estimated at 16 per cent of the total within the industry in 1982 and this proportion has risen to 20 per cent by end 1988.

The relatively low profitability and increased susceptibility to bankruptcy is partly explained by the small size of the typical building materials producing firms. Only two firms had assets in the Rmb 300 to Rmb 500 million range - they were the largest firms in the industry in 1987 and together they accounted for less than 2 per cent of gross output value in that year. As against this, 63 enterprises with assets in the range of Rmb 50 to Rmb 100 million accounted for 54 per cent of gross output value. Building materials producers thus cannot take advantage of scale economies and horizontal integration of units and rationalization of production has proceeded extremely slowly. Building materials industry has thus yet to fully incorporate the organizational restructuring that has been under way since 1984.

Selected performance indicators of State-owned construction enterprises are presented in Table IV.13. Fourteen out of 24 performance indicators show subdued or declining performance levels in 1990. Further expansion of the building materials industry requires significant improvement in the levels of factor productivity. An important development in recent years has been the gradual introduction of new building materials such as frame structures and lightweight board into construction processes. The labour productivity of construction workers using these new materials is significantly higher than those of workers using brick and concrete structures. The use of the new building materials is energy conserving. Demand for this type of material significantly exceeds its supply. New units for producing such construction materials have been established in many cities. Production capacity is about four million square metres of floor space.

Table IV.13. Selected performance indicators of State-owned construction enterprises, 1987-1990

Indicators	Unit	1987	1988	1989	1990
Gross output value	1,000,000,000 yuan	66.01	77.70	87.86	93.52
Construction output value	1,000,000,000 yuan	62.78	73.72	82.82	89.03
Overall labour productivity in terms of gross output value	yuan/person	10345	12168	13820	14509
Overall labour productivity in terms of construction output value	yuan/person	9839	11545	13027	13812
Average number of total staff and worker	1,000	6380.8	6385.4	6357.4	6445.5
Excluding other personnel[a/]	1,000	5957.2	5889.8	5894.5	5840.5
Floor space of buildings under construction	1,000,000 sq m	211.53	222.97	213.99	203.03
Floor space of building completed	1,000,000 sq m	92.09	88.99	91.43	93.62
Residential buildings	1,000,000 sq m	36.62	35.33	35.16	38.39
Completion rate of floor space of buildings	percentage	43.5	39.9	42.7	46.1
Rate of fine-quality projects	percentage	58.0	51.5	46.7	44.0
Net value of mechanical equipment possessed at year-end	1,000,000,000 yuan	18.17	18.00	19.09	20.35
Per capita machinery	yuan/person	2939	2887	3105	3277
Per capita machinery capacity	kW/person	4.6	4.5	4.8	4.9
Net output value	1,000,000,000 yuan	19.16	22.84	25.38	27.10
Reduction of construction cost	1,000,000,000 yuan	2.45	2.14	1.32	0.34
Reduction rate in construction cost	percentage	4.2	3.2	1.8	0.4
Total profits	1,000,000,000 yuan	2.80	3.18	2.64	1.66
Ratio of profits to gross output value	percentage	4.2	4.1	3.0	1.8
Average balance of quota circulating fund	1,000,000,000 yuan	21.86	24.86	29.17	33.96
Ratio of target circulating fund to gross output value	percentage	33.1	32.0	33.2	36.3
Original value of fixed assets (year-end)	1,000,000,000 yuan	43.37	45.49	51.18	55.40
Net value of fixed assets (year-end)	1,000,000,000 yuan	31.99	32.97	36.24	38.23
Ratio of profits to total funds	percentage	5.2	5.5	4.0	2.3

Source: State Statistical Bureau of China, *China Statistical Yearbook* (various issues).

a/ Other personnel refers to persons who are engaged in farm and sideline production, working abroad in foreign-aid projects, on leave for study or training over six months, on leave due to sickness or injury over six months, and those sent to work in other units.

Prospects

Rapid growth of the building materials industry is needed to meet existing demand, particularly for the development of the new towns all over the country. Shortages are most acute in the case of plate glass and ceramic fixtures. Expansion of production must be accompanied by reduction in costs - particularly energy costs per unit of production. These remain high by international standards. Production costs have risen in recent years. An improvement in performance requires greater emphasis on the process of vertical and horizontal integration. There is also a need for demand management. The product mix must be designed to cater to demand which has priority within the national development and investment plan. During the 1980s, imbalances occurred as the very rapid growth of luxury offices, hotels and foreigner-occupied houses created shortages of building material supplies in key development projects.

There is a need to widen the application of foreign technology in order to facilitate the rapid expansion of the building materials industry in China. Attention is being focused by the State Building Materials Bureau on the injection of foreign advanced technology. It encourages the imports of key equipment for production lines for stone materials, asbestos, and cement software technology. Scientific managerial expertise is also being increasingly bought to upgrade old factories. Utilization of foreign funds for developing non-metallic mineral products for export is also envisaged.

F. MACHINE TOOLS AND MACHINE BUILDING: NEED FOR A COMPETITIVE ENVIRONMENT

Machine Tools

An overview

Following decades of tardy development, the machine tool industry in China started to expand in the 1950s concomitantly with increased emphasis placed on heavy industry. During 1953-1957 the machine tool industry absorbed 18 per cent of the investment under the first Ministry of Machine Building. Aided by the massive material and technical assistance provided by the USSR and east European countries, the number of specialized machine tool factories expanded rapidly. A large number of small-scale plants were also set up. By 1960, China was almost self-sufficient in commonly used machine tools.

As the spectrum of machine tool production expanded, China produced over 550 types of machine tools by the year 1965. Having suffered a downturn in output during 1967-1968, the machine tool industry continued to increase the variety of machine tools, and by 1972 over 800 types of commonly used machine tools and 1,100 varieties of specialized machine tools were turned out. When this subsector rebounded in the 1970s, considerable qualitative improvements took place. However, the greater part of the commonly used machine tools were of 1950s or 1960s vintage.

By 1987, there were 228 State-owned metal cutting machine tool factories,[17/] employing 307,370 persons. Of the total number of employees 25,582 were engineers and technicians. The gross industrial output value of these enterprises stood at Rmb 2.6 billion in 1987. With the capacity to produce over 2,200 varieties of products, these enterprises turned out 120,000 units in 1987. Data pertaining to the production of high precision and numerically controlled machine tools in recent years show subdued production trends. Of the total of 134,500 units of metal-cutting machine tools in 1990, large-scale units represented 2,295 (see Annex Table A-19). The production of high precision equipment rose marginally while that of numerically controlled machine tools fell from 2,742 units in 1989 to 2,634 units in 1990.

China has a large number of small firms and a small number of exceptionally large firms. The proliferation of small machine tool-making firms was due largely to the rigidity of administrative planning which called for local self-sufficiency in these key investment goods. In the absence of competition, these firms in different localities produced old vintage, low value tools. There was a growing tendency among machine tool producers to use their new freedom to determine the allocation of productive resources. As a result, there was a tendency to shift a part of their productive capacity from machine tools to directly producing light industrial machinery, such as knitting machinery, cigarette rolling machines, plastic pressing machines, etc. A stagnation in the production of machine tools occurred at a time when the industrial sector received fresh impulses to grow rapidly in the 1980s. As the horizontal links through market mechanism started to emerge, new patterns of machine tool use surfaced.

Exports and imports

During 1956-1978, China exported 51,000 machine tools, representing 2 per cent of the national output which accounted for 0.26 per cent of the value of world machine tool exports. By 1987, China's share in world machine tool imports rose to 0.45 per cent. The legacy of economic isolation resulted in little exchange of technical knowledge, and technical progress embodied in imported machine tools contributed little. In the 1980s, China's imports of advanced machine tool technology rose significantly. This occurred largely through joint venture agreements. China was also purchasing foreign blueprints. In contrast to the low value of China's machine tool exports, the value of imported items was higher all through (see Table IV.14). In 1990, the average value of China's machine tool exports stood at $355, compared with $11,258 for imports, reflecting the higher value of advanced machine tools imported by China. The lower value of machine tool exports is due to low quality products and poor after-sale service. Competition among domestic machine tool producers and direct interaction between producers and consumers will facilitate the development of the country's machine tool industry, which could successfully integrate imported technology into indigenous production.

Table IV.14. Exports and imports of machine tools, 1981, 1985 and 1988-1990

	1981	1985	1988	1989	1990
Exports:					
Number of units	33344	17198	433218	603575	624898
Value (million $)	57.9	16.7	131.9	188.3	211.7
Average value ($)	1736	971	304	312	355
Imports:					
Number of units	2534	13006	44489	41630	43773
Value (million $)	47.7	137.1	519	486.3	492.8
Average value ($)	18824	10541	11666	11681	11258

Source: State Statistical Bureau of China, *China Statistical Yearbook* (various issues).

Taiwan Province successfully followed the path taken by Japan in the 1960s, reversing the machine tool industry on the basis of imported models and then making incremental improvements independently. Thus, Taiwan Province stands as a parallel success story. As in the case of Taiwan Province, China is able to purchase technology through licensing agreements. A conducive institutional framework, careful scrutiny of competitive products and constant feedback from domestic and foreign customers were the major determinants of technical progress in machine tool industry both in Taiwan Province and Japan.

Prospects

A poor record of indigenous innovation does not necessarily constitute a constraint on the shift towards newer vintages based on imported technology. Nor does it imply that it is desirable to halt the production of traditional machine tools in China. A large reservoir of human skills produces and operates machine tools in China. Given the lack of technicians with advanced training in handling modern tools, human skills in traditional machine tools could be augmented for repairing and maintaining them whenever the traditional technology is deemed appropriate. Because of the proliferation of a large number of small-scale enterprises with limited ability to fully use modern machine tools, even with the increased accent on modern technology, a large proportion of machine tool production is likely to continue on traditional lines.

Creating a competitive environment is pivotal to enhancing efficiency. Although China is able to produce a wide range of machine tools the technical capabilities of the Chinese machine tool industry are still far below the level of industrialized and Newly Industrializing Economies. According to rough estimates, the cutting speed of lathes in China is 80-100 metres per minute, compared with an average of 200 metres per minute for other countries. The cutting speed of grinders in China is around 35 metres per second compared with an average of 60 metres per second for other countries. Similar trends have been estimated for the speeds in finishing gear-lobbing machines and automatic lining of combined lathes. The relatively low level of technical efficiency in Chinese machine tool industry is attributed largely to the lack of a competitive environment. A breakthrough in machine tool industry calls for the erection of competitive pressure in order to enhance technical efficiency.

Fig.IV.6. Physical output of selected machine tools, 1984 and 1990 (Thousand units)

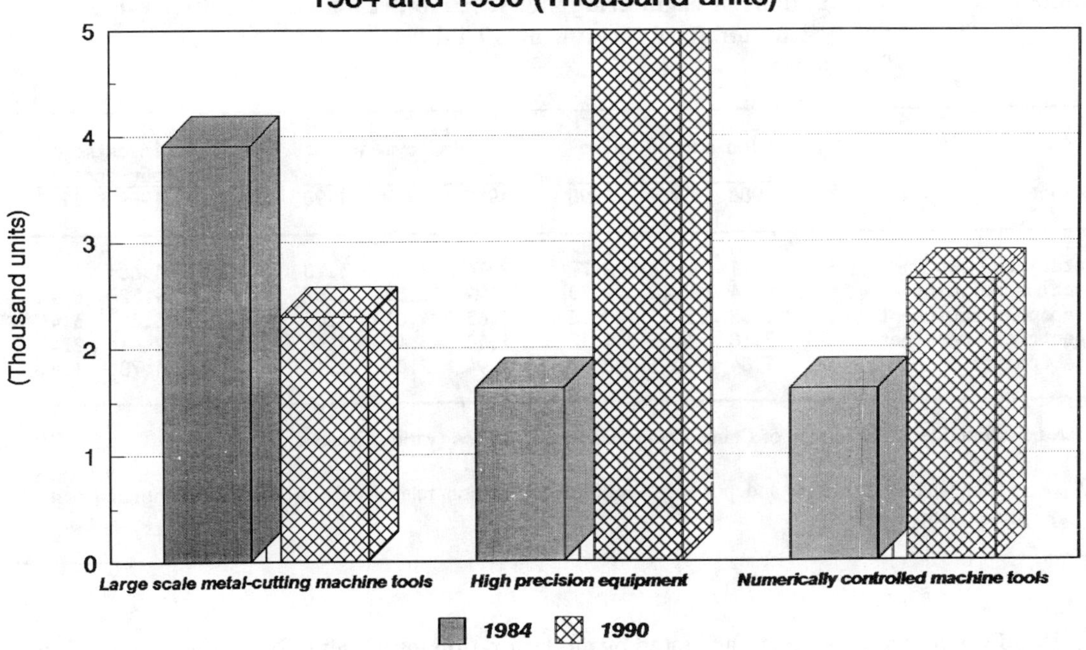

Machine building: Quest for technological upgrading

The machine building industry is the largest branch of the engineering sector (see Table IV.15). It accounts for almost 50 per cent of net output and about 60 per cent of the employment generated within the sector. However, its share in investment is much lower, about 25 per cent. The transport equipment branch which produces about a quarter of the net value generated by the machine building sector has a significantly higher investment share. This reflects both the emphasis the government has placed in recent years on the rapid development of transport infrastructure and the fact that the rate of return on investment (measured in terms of profits and taxes as a ratio of the value of fixed assets and circulating capital) has in recent years been lower in this branch than in other engineering branches.[18/] China's machine building industry encompasses among other things the production of farm machinery, metallurgical equipment, mining equipment, chemical industry equipment, lift and transportation equipment, printing machines, power equipment, transportation and transmission lines and electrical appliances, instruments and metres, packaging equipment and food processing machinery.

The typical machine building enterprise is modest in size - only three firms have assets in the Rmb 500 million - Rmb 1 billion range. These accounted for just 1.6 per cent of the gross value of output produced by the machine building sector in 1986. As against this the 169 enterprises in the Rmb 50-100 million range and the 1,084 enterprises in the Rmb 10-30 million range produced 21.0 per cent and 36.6 per cent of the gross value of branch output in 1986 respectively. The industry is thus not structured to efficiently exploit existing economies of scale and horizontal integration remains an important necessity. There has been significant growth of non-State-owned enterprises since 1984 - they accounted for about 30 per cent of employment within the branch during 1985-1987. The output value share is however likely to be limited - no detailed statistics are available. The predominance of State units is indirectly reflected in the fact that roughly 85 per cent of the gross value of output produced within the machine building sector is classified as the product of heavy industry in Chinese statistical sources.

Table IV.15. Contributions of engineering industries to net industrial output value[a/], employment and fixed investment, 1988-1990
(Percentage)

Branch	Net output value			Employment			Fixed investment (year-end)		
	1988	1989	1990	1988	1989	1990	1988	1989	1990
Metal products	1.71	2.78	2.77	2.74	3.76	3.70	3.87	1.68	1.67
Machine building	10.24	10.60	9.50	10.99	12.86	12.67	13.11	9.52	8.98
Transport equipment	3.82	3.55	3.68	3.65	4.01	4.05	4.03	3.61	3.46
Electrical machinery	2.38	4.55	4.12	4.10	3.47	3.48	3.50	2.41	2.47
Electronics	1.84	2.91	2.87	2.94	2.06	2.12	2.02	1.78	1.84

Source: State Statistical Bureau of China, *China Statistical Yearbook* (various issues).

a/ Manufacturing, mining and power generation. Information relates only to enterprises with independent accounting systems.

Project diversification has been facilitated by modernization investment. New products developed in recent years include continuous casting systems for steel plants using 1970s technology with an annual capacity of 105,000 to 300,000 tons of square blooms, high pressure gasifiers with hourly capacity of about 15,000 cubic metres of coal gas, poultry farming installations such as incubators and battery-house control devices, energy-saving devices such as parametric equipment, compressors and transformers, digital process control machine tools, etc. Beginning in 1983, international standards are being applied to a wide range of machine products, particularly those designed as components of energy-saving systems. Improving product quality has been facilitated by increased access to foreign technology. Further rapid growth in technology imports is envisaged as an indispensable basis for modernizing the machine building industry.

The machine building industry is primarily domestic demand oriented. Only about 300 out of a total of over 3,000 products are currently exported. Exports include, among others, bearings, electric motors, diesel engines and digital process control grinders. Complete machine building plants and transmission and transformation sub-stations have also been exported to some developing countries. Thailand, the Philippines and Pakistan have imported machine building technology from China in recent years.

Since 1984 there has been almost a continuous process of re-organization of production and management structures within the machine building sector. Currently, as many as 40 departments are involved in the machinery and electronics industry at the central, provincial and municipal levels including the Ministry of Machinery and Electronics, ministries and departments of the main sectors utilizing the equipment, military industries and township and village industries. Of these, the Ministry of Machinery and Electronics is the most important accounting for 11.4 per cent of all enterprises, 45 per cent of employment, 57 per cent of fixed assets and 50 per cent of total output in 1989. In 1986, the Ministry established the State Commission for the Machine Building Industry.

In 1987 and 1988, progress was made in designing and producing equipment for many key projects in China. The State Commission completed 80 per cent of the design for blast-furnace equipment to be installed in the second-phase construction of the Baoshan Iron and Steel Works in Shanghai. The equipment for continuous cold sheet mill, which the Commission built in co-operation with other factories, was installed in the works. Some progress was also made in developing key military equipment, large-sized mining machinery, railway equipment and equipment for the building industry, extra high voltage power transmission and transformation equipment, machinery for coal chemical engineering plants, equipment for 300,000 tons a year ethylene plants and large-sized equipment for fertilizer works. The enterprises under the Commission completed 98.9 per cent of their capital construction targets, and 61.1 per cent of the fixed assets built were put to use. In 1987, 36 projects and 113 technical upgrading projects were completed and put into operation.

The enterprises under the State Commission for Machine Building Industry doubled their total output value in 1987 to reach Rmb 90.09 billion, fulfilling their plans two years ahead of time. The sales amounted to Rmb 93.83 billion, both showing an increase over the previous year.

The generating equipment manufactured in 1987 has a total capacity of 9.2 million kW, an increase of 35 per cent over 1986, while main engines and accessory equipment were produced to equip generating units with a total capacity of six million kW. Chinese-made 300,000 kW and 320,000 kW generating sets have been installed and begun to operate.

A 300,000 kW thermal generating unit, made in China with technology imported from the United States, was put into operation at the Shiheng Power station, Shandong Province while a 600,000 kW thermal generating set made with United States technology was installed in 1988. The technical parameters and other related indexes of the two sets are up to advanced world levels.

The 500,000 volt transmission line erected in Liaoning Province, the country's first with sole Chinese equipment, passed State appraisal in 1987. The power transmission equipment for another 500,000 volt transmission line developed with imported technology operates successfully. Installed in the Gezhouba Dam, the 500,000 volt transmission equipment, 70 per cent of which was domestically made, is up to the international levels of the mid-1970s.

By the end of 1987, China's capacity for railway locomotive production ranked second in the world (next to the Soviet Union). More specifically, the country had a capacity to turn out a diesel locomotive in 12 hours, an electric locomotive in 33 hours and a passenger carriage in 2.6 hours.

The country's machine building industry has imported 1,200 technological items over the past few years, making up for one third of the total software technological imports. The imported technology has been introduced to 1,000 key enterprises and been incorporated in more than 10,000 varieties of products. There were 119,000 rural machine building enterprises in the country by the end of 1987, employing 4.64 million workers and turning out a total of Rmb 44.38 billion worth of products. Their output value accounted for 19.37 per cent of that of the national machine building industry.

Another important segment of the machine building industry is farm machinery. More than 3,200 kinds of farm machinery and equipment are now produced in 2,266 factories in the country. Farm machines in China had a total horse-power of 250 million kW at year-end in 1987, nearly three times that of a decade ago. The number of farm tractors increased 3.3 times to reach 23 million in the decade ending 1987. Tracks used in the farms number 550,000, registering a more than sevenfold increase in this period. There were also sharp increases in tractor-drawn implements, trailers, farm produce processing machinery and other equipment.

Chinese peasants can now own farm machines or contract them from the collectives, thanks to the institution of the contract system in the rural areas. The peasants have been spending large sums of money on purchasing farm machines annually over the past few years, and the number of farm machines privately owned accounted for over 80 per cent of the national total in 1988.

The setting up of repair centres and other service stations has kept pace with the increase in the number of agricultural machines. At present there are such service centres that not only sell farm machines or repair them for peasants, but also help in training them on the use of farm equipment of all kinds.

More than 20 important technological items were introduced into farm machinery manufacturing during 1985-1988. This, coupled with stricter standards set for the farm machinery industry, has resulted in the turning out of better machinery. However, tests conducted on 500,000 tractors in different parts of China show that most of the farm tractors on Chinese farms have a technical level comparable only to that of machines used on farms in the industrialized countries in the 1950s or 1960s. Some 30-40 per cent of existing tractors are old, very low in efficiency and consume 28 per cent more fuel than modern machines.

Despite impressive progress made since 1984, the machine building industry remains beset with many problems. Profitability within the sector remains low and this is especially true of the large State-owned enterprises. The very rapid growth of investment in 1988 was largely financed by bank borrowing. Moreover, inflationary pressure has meant that working capital requirements have risen very rapidly - sometimes to over 80 per cent of annual capital requirements.

The acceptance rate from the sample check made on the products, the quality of which was monitored by the State, had been fluctuating at about 50 per cent. The quality of some products has dropped. For example, sample checks of the State monitoring nature were performed on 131 excellent quality products in 1986. Out of these 131 excellent quality products, 43 were downgraded to qualified product states or even downgraded to unqualified ones, accounting for 32.8 per cent of the total.

Based on the statistics gathered from the Second Industry Census, analysis was made on 2,500 pieces of major industrial equipment in 91 categories in the key enterprises that were affiliated to the Ministry of Machine Building Industry. The result showed that only 13.7 per cent of this equipment was of the advanced international level. Equipment of the domestic backward level that was obsolete and needed to be updated accounted for 35.9 per cent and equipment of the general domestic level accounted for 50.4 per cent. These two types of equipment together accounted for 86.3 per cent of the total equipment. The reforms and the development of market relations do not seem to have significantly transformed this aspect of machinery production in China.

G. AUTOMOBILES: GROWING IMPORTANCE OF JOINT VENTURES

The evolution of the industry

China's First Auto Works (FAW) came on-stream in 1956 (Changcun in Jilun Province) but reached full production in 1965. Based on Soviet technology, a four-ton truck with a 110 hp engine was produced. The origin of its technology could be traced to a Ford-built vehicle of the 1930s vintage. The Second Auto Works (SAW) with facilities to produce 4.5-5.0 ton range trucks became operational at Shiyan in Hubei in the early 1970s. Around 95 per cent of the components used were of Chinese origin. Each of these two automobile complexes are capable of turning out 100,000 trucks annually. In the mid-1980s, over 120 small assembly operations were scattered throughout China. By the late 1980s, 23 provinces had facilities for assembling automobiles. Of the 472,000 automobiles produced in 1987, 45 per cent came from the First and Second Auto Works. China currently produces about 600,000 automobiles, comprising more than 50 kinds of vehicles including passenger cars, light, medium and heavy trucks, buses and other vehicles. The proportion of cars, buses and light trucks increased in the 1980s while that of heavy vehicles faltered (see Annex Table A-20).

To improve the industry's management, the State Council established in June 1987 the China National Automotive Industry Corporation (CNAIC) which serves as a bridge between enterprises and government by providing services to both. The corporation's main tasks are: studying the developments of the industry at home and abroad in order to make policies about development strategy and planning of China's industry; organizing the drafting of regulations and standards for

Fig.IV.7. Production of automobiles, 1978 - 1989, selected years ('000 units)

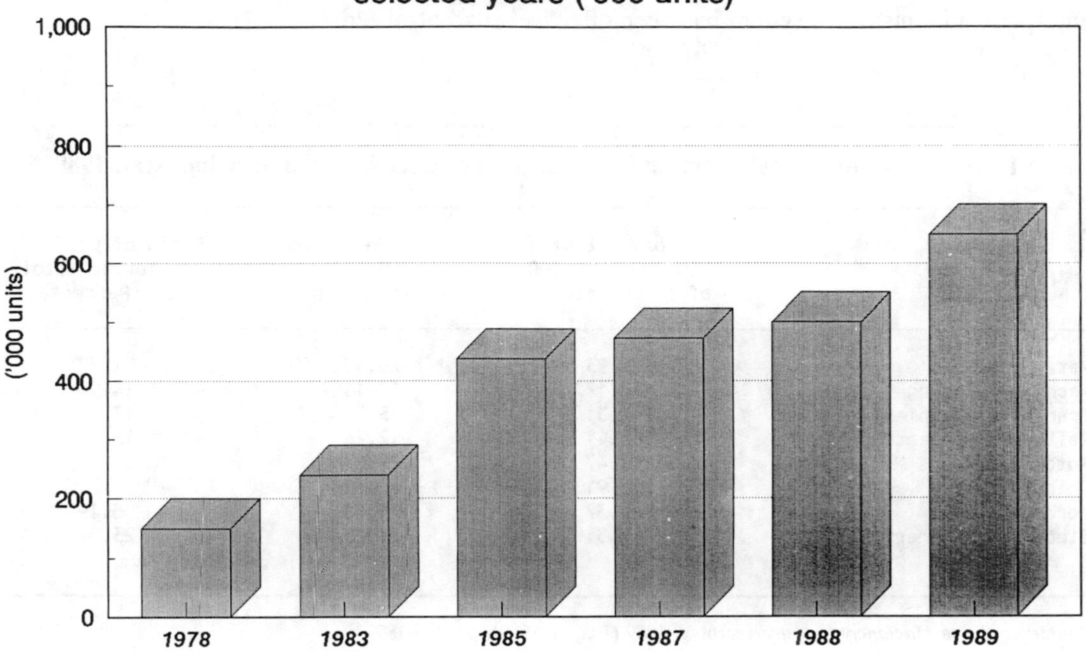

the industry, granting production licences to enterprises and supervising the quality of products; and co-ordinating the industry's production, construction, scientific research, technological import and export policies.

The corporation has developed an automotive development strategy whereby the leading role is assigned to the First Auto Works and Second Auto Works in combination with smaller local and defence enterprises. Specifically, the First Auto Works, concentrating on high- and medium-grade cars equipped with 2.2 and 2.5 litre engines, will have a capacity of manufacturing 30,000 cars annually by 1991. This will rise to 150,000 a year eventually. Second Auto Works, which concentrates on general purpose cars equipped with 1 and 1.6 litre engines, will turn out 150,000 cars annually after the first phase of the construction project is completed and then turn out 300,000 cars a year after the second phase of construction.

During the reform decade China has actively encouraged joint venture collaboration within the automobile industry. Major joint ventures include the Beijing Jeep Corporation producing Cherokee jeeps and Shanghai Volkswagen Automobile Company which produces Sautand cars. The automotive exports have not yet integrated China into the international industry circuit but considerable potential exists for the rapid expansion of Chinese automotive exports.

China plans to increase automobile production from about 580,000 units annually in 1990 to 1,700,000 units annually by 2000. There is also a concern to diversify production and expand regional integration. Employment generated by the auto industry currently amounts to a little over 1.5 million persons - less than 10 per cent of whom are technical personnel.

Performance

Despite the predominance of the two major plants (FAW and SAW), the total number of automobile and motor-cycle manufactures has grown rapidly in recent years and now exceeds 4,000. At end 1987 gross industrial output value of these enterprises totalled Rmb 21,866 million. Growth of output value has exceeded 20 per cent annually since 1986. The financial performance of automobile companies has been superior to that of all other segments of the machinery industry (see Table IV.16). The relatively higher profitability levels of automobiles reflect the greater level of enterprise autonomy that has been made possible by managerial re-organization, horizontal integration and the establishment of the China National Automotive Industry Corporation. The enterprise administrative system has been effectively decentralized.

Table IV.16. Automobiles' superior financial performance in machinery industry, 1988

	Gross output value per Rmb 100 of fixed assets	Profit and taxes per Rmb 100 of fixed assets	Ratio of profit and tax to total capital (Percentage)
Metal products	175.23	26.48	21.72
Machine building	120.17	17.14	14.58
Transport equipment	117.31	15.28	13.30
Railway equipment	105.41	12.76	13.23
Automobiles	181.47	29.07	25.23
Ships	80.91	2.61	1.74
Aeroplane	67.37	6.88	8.19
Electric machinery	160.51	31.85	23.07

Source: *China Machinery Industry Yearbook 1989* (Beijing 1990), pp. 187-187.

In contrast to the relatively better financial performance of the automobile industry, capacity utilization remained far below the optimal level in the late 1980s, revealing a limited degree of specialization and the geographically localized nature of supplier networks. Low quality parts supplied by subcontractors, the weakness of co-operative information sharing arrangements, the slow pace of technological advance and the persistence of small-scale assemblers and component manufacturers contribute to insignificant improvements on products. Several models of the vehicles manufactured in the late 1980s remained unchanged for over three decades. Twenty-year old equipment is still in use, and production tooling is out of date.

Joint ventures

Table IV.17 presents the list of joint ventures in the production of vehicles. The joint ventures are under strong pressure to source most parts locally by the late-1980s. The initial experience of Beijing Jeep, Volkswagen and Peugeot suggests that local design skills are not encouraging, and producers lack the machinery needed to make licensed components. The defense factories that are better equipped have no experience in producing large batches of auto parts; there remain other constraints that tend to slow the emergence of an effective auto components network in China; and co-operation between the various producers is rather limited.

Table IV.17. Joint ventures in automobile industry, 1988

Companies	Type of vehicle
Guangzhou Automotive/Peugeot	Car/van
Shanghai Auto/Volkswagen	Car a/
Tianjin/Daihatsu	Mini van
Chongqing Auto/Isuzu	1.5-1.75 tons
Sichuan Motor/Berliet	Heavy truck
Jinan Auto/Steyr	Heavy truck
Beijing/Jeep	Jeep
North Industrial Corporation of Beijing (NORINCO)/Benz b/	Truck
FAW/Audi-VW/Benz c/	Car

Source: "Daimler to make heavy trucks in China", *Financial Times*, October 28, 1988.

a/ 15,000 Santana cars were produced in 1988, and 17,000 were planned for 1989.

b/ Trucks in the 16-36 ton range will be assembled, initially from SKD kits using locally manufactured Deutz diesels and ZF gear boxes. Production is expected to reach 6,000 per annum by the mid-1990s all from the Baotou plant.

c/ FAW had been assembling a few Mercedes Benz cars since 1987 and could expand this operation. It is also reviewing the possibility of reengining the Red Flag limousine with a Mercedes V-8.

Efforts to form the second car assembly joint venture with Volkswagen of Germany are under way. This new venture, FAW-Volkswagen Automobile Company, is aimed at building 150,000 cars a year by 1996. Sixty per cent of the stake will be owned by the First Automobile Works. The initial equity capital of this new venture is estimated at $405 million. The plan is to assemble Volkswagen Golf-Jetta cars for the domestic market and export, particularly to countries in South-east Asia. However, initial production of around 17,000 cars will be exclusively for the domestic market. Exports are scheduled to commence in 1994.

As part of an earlier licensing and technology transfer agreement with Volkswagen of Germany, the First Automobile Works has been assembling the Audi 100 model car since 1989. Its full capacity is expected to reach 30,000 a year in 1995. The new joint venture is expected to achieve a local content level of 65 per cent in 1994.

Prospects: new challenges

The development of the auto industry poses a set of problems. First, the infrastructural investment necessary to sustain a rapid auto production expansion is sizeable. In particular, a major development of the road and construction programme is required. Secondly, the relative growth of cars against public transport poses both social and environmental issues. Thirdly, while technology has been achieved slowly and at great cost, there has been a growing dependence on foreign suppliers and joint venture partners who have been reluctant to source locally and even more reluctant to promote production for export. China remains eventually an exporter of auto components and the growth of the auto industry is unlikely to lead to a significant increase in foreign exchange earnings in the near future. The further development of the auto industry thus poses hard policy choices for Chinese decision-makers. It is probable that top priority will be accorded to rationalization and horizontal integration of the national industry so as to increase its capacity to adapt imported technology over the medium term.

Car makers and component suppliers face new developments in the international market due to heightened competition. New challenges stem from the build-up of Japanese "transplant" car production in the hitherto protected European Community, the impact of the Single European Market on car pricing, the opening up of eastern European markets, and costly demand from environmentalists. While Europe seems likely to emerge as the main area of competition for the world automobile industry in the 1990s, an overcapacity problem is likely to surface if the Japanese continue to increase their exports. Japanese car makers have captured over 30 per cent of the United States market where leading car makers are operating at a loss. The state of the United States auto manufacturers serves as a warning to European manufacturers who try to close the competitive gap. These trends have far-reaching implications for China. Most of the joint ventures in vehicle production in China are with European firms. China's automobile industry will need to gear up for the new challenges their joint venture partners now face on the world market.

H. ELECTRONICS AND ELECTRIC APPLIANCES: IN SEARCH OF HIGH-TECH

An overview

The electronics industry is the smallest but among the fastest growing segments of the engineering branch in China. It accounts for roughly 10 per cent of the engineering industry's net output value and 1.3 per cent of fixed investment. Modernization investment in the electronics branch, however, represents about 20 per cent of the total modernization expenditure undertaken by the machine building industry.

Roughly 50 per cent of the gross output value of the electronics industry is categorized under light industry production. The electronics industry is mainly concentrated in the coastal provinces - Beijing, Tianyian, Shanghai, Jiangsu and Guangdong together accounted for about 60 per cent of its gross industrial value in the late 1980s. The only non-coastal province with a sizeable electronics industry is Sichuan. Despite the relatively rapid growth of the electronics industry in recent years, the rate of return in this industry is modest - the surplus to assets ratio was 14.1 per cent in electronics, compared to 18.1 per cent in electrical equipment and 17.2 per cent in metal products during 1985-1986. However, labour productivity has grown more rapidly in electronics than in any other branch of the machine building industry. The electronics industry is relatively more concentrated than the machine building branches. Eight firms with assets in excess of Rmb 300 million account for roughly a fifth of gross output value. The electronics industry is thus better placed to take advantage of the economies of scale, and less beset with the type of problems which make horizontal integration of firms difficult.

Enterprise re-organization has proceeded rapidly within the electronics subsector. Over 170 enterprises belonging to the Ministry of Machinery and Electronics Industries have been decentralized since 1982. For these enterprises, management and planning have been decentralized, and during the 1982-1988 period the Ministry only gave major directives. A number of larger enterprises have become industrial groups through amalgamation and merger. Decentralized management has led to better organizational structure of enterprises. It has also stimulated scientific research, education and training and speeded up commercialization of scientific research.

The performance of the electronics industry in terms of the growth in output of many products has been quite impressive. From 1978 to 1987, the average annual growth rate of the electronics industry exceeded 20 per cent. During 1988, a 30 per cent increase in gross output was achieved.

China's electrical appliances industry produces a wide range of household items, such as refrigerators, air conditioning, cleaning, ventilating and heating appliances as well as kitchen, cosmetic and health-care utensils and a variety of accessories for special purposes. In the late 1980s, there were around 1,000 enterprises turning out a gross output value of Rmb 24 billion. In 1989, the industry produced 6.7 million refrigerators, 8.3 million washing machines, 49.9 million electric fans and 14.0 million electric irons. Currently, China ranks first in the world production of washing machines, electric fans, electric rice cookers and electric irons, and second in the production of refrigerators.[19/]

In the urban areas on an average 100 families own around 16 refrigerators, while in rural areas the corresponding figure stands at 2.5 for 100 families. With an improvement of power supply in the rural areas, a greater demand for refrigerators is expected. There is a relatively higher possession rate of washing machines; 76 per 100 households in urban areas and 11 in rural areas. The market situation for electric fans is much better than the use of refrigerators and washing machines. There are 143 electric fans for every 100 families in urban areas and 57 in rural areas. According to rough estimates, rural demands for electric fans currently account for 64 per cent of the total demand.

Leading products: diversified mass production

The most successful product launched by the electronics industry in recent years has been colour television. In 1987, the enterprise under the Ministry of Machinery and Electronics Industries turned out 16.8 million television sets (including six million colour sets), 22.5 per cent more than in 1986. They produced 15 million tape recorders (an increase of 15.5 per cent), 59,000 micro-computers (51 per cent more) and 78 million integrated circuit chips (70 per cent more). The output of washing machines, refrigerators, electric organs, telecommunications and broadcasting equipment, satellite monitoring stations, electronics components and parts were much greater than in previous years.

The quality of electronics products has improved steadily. More than 230 top quality products are currently made to international standards. Of all television sets made in 1989, 27 brands of colour sets won top quality awards. The trouble-free working time of colour sets is over 20,000 hours. New achievements have been made in R & D. In 1987, 66 research projects won State scientific and technological advance awards. Among them is the "Hua Guang" computer-laser Chinese characters composing system, which is up to the international advanced level.

Large-scale integrated circuits, colour tubes, programme controlled exchangers, video-cassette recorders and computers were listed as the five major projects under the Seventh Five-Year Plan for the electronics industry. Progress was made in these projects during 1987-1989. Construction began on a number of colour television tube projects. Negotiations are under way with foreign firms on large-scale integrated circuit, programme controlled exchanger and video cassette recorder projects. Micro-computers of up to 16 digits are sold extensively on the domestic market.

Television sets: accent on colour TV's

Buoyed by strong domestic demand, the colour television industry has grown rapidly in recent years. Between 1977 and 1986, over 50 new colour television firms came on-stream. Table IV.18 shows the household penetration[20/] of colour and black and white televisions. The percentage of black and white televisions per household rose from 0.6 per cent in 1976 to 40.9 per cent in 1987, and that of colour televisions rose from 0.01 to 10.0 during the same period. Projected demands for television sets shows a stagnant market for black and white televisions and a significant increase in the demand for colour televisions (see Table IV.19). China constitutes the second largest market for television sets after the United States.

According to the Ministry of Machinery and Electronic Industries, the top ten manufacturers accounted for around 50 per cent of total production of colour televisions in the mid-1980s, and the top five accounted for about 30 per cent of production. Despite economies of scale, the production of colour televisions is generally characterized by very high costs due largely to huge idle capacity and the high costs of inputs and working capital. Given the strong domestic demand for colour televisions in the huge internal market, domestic sales remain very profitable.

In the mid-1980s, unit costs of materials for Chinese manufacturers were significantly higher than the material costs of international manufacturers. This was attributed to a larger share of imports in material inputs, higher foreign exchange rates paid by importers of components and tight world markets for these components in the mid-1980s. Retail prices for colour televisions in China were estimated to be 65 per cent higher than in the world market in the 1980s.

Fig.IV.8. Production of selected electric and electronic items, 1980 and 1990 (Million units)

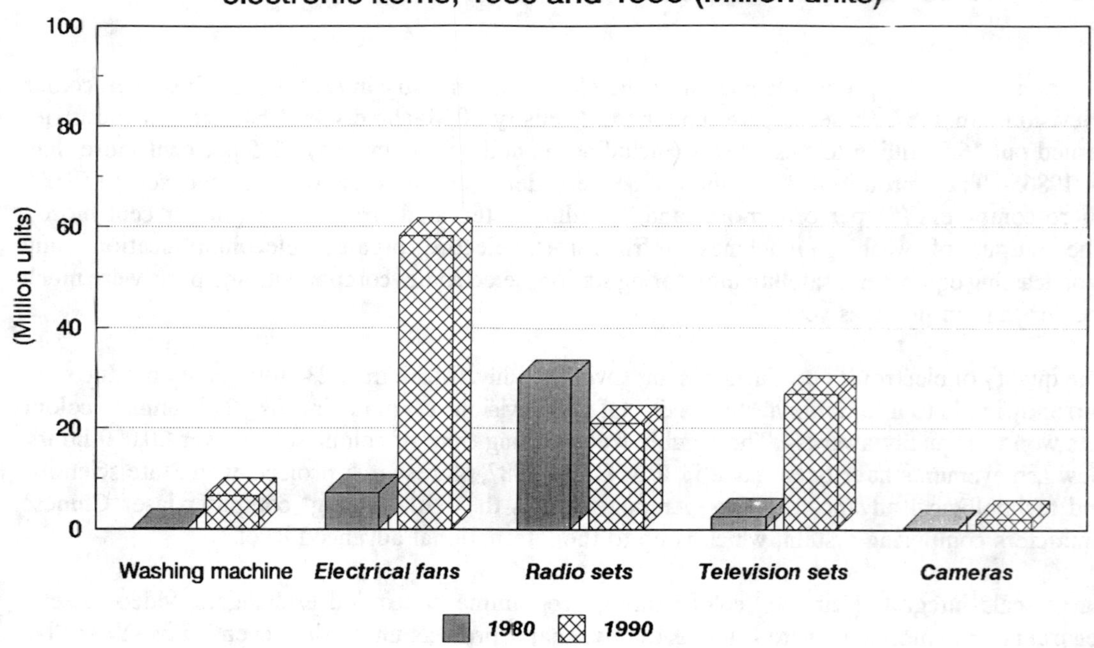

Table IV.18. Household penetration[a] of television sets in China, 1976-1987, selected years

	1976	1980	1984	1987
Number of households (million)	162.0	183.2	207.0	220.0
Number of black and white TV sets (million)	1.0	8.7	40.2	89.9
Number of colour TV sets (million)	0.01	0.4	10.9	22.0
Household penetration (percentage)				
Black and white TV sets	0.6	4.8	19.4	40.9
Colour TV sets	0.01	0.2	5.3	10.0

Source: Ministry of Machinery and Electronics Industries.

a/ Household penetration is defined as the number of television sets per household, expressed as a percentage.

Table IV.19. Projected demand[a] for television sets, 1990 and 1995
(In million units)

	1990	1995
Black and white television sets	12	12
Colour television sets	14	20.5

Source: China Electronic Engineering Consulting Company.

a/ Projected at the beginning of the Seventh Five-Year Plan (1986-1990).

Computers: great strides

The 1980s witnessed rapid development of the computer industry in China, with a growing emphasis on the production and use of micro- and mini-computers. Although most of the production was based largely on imported devices and components, basic research and product development was under way.[21] Some of the products have reached international standards and have been widely used. The Five-Stroke Coding System using the Chinese radical periodic table with the aid of 26 keys can handle thousands of Chinese characters. The Beijing Union University developed the Laser Series Chinese-language information processing module for non-screen language editing and graphic illustrations. Some unique language editing and information processing software, developed in 1989, is used by over 20 newspapers, magazines and presses.

China also successfully produced the Galaxy and Shenghou computers capable of 100 million operations per second, and other high-speed computers with millions of vector operations - compatible with international standards. The Shenghou micro-computer is capable of transferring data to large-capacity computers. The mini-computers ST-286H and Taiji were manufactured by Beijing Xintong Computer Company which won an international tender in a competition with over 30 world-famous computer companies.

Thus, in recent years, China's computer industry has achieved great strides in both hardware and software. By the beginning of 1991, China had produced 5,000 large-, medium- and small-sized computers, 160,000 micro-computers, 80,000 Zhonghua-branch educational computers and 220,000 peripherals. China's computer industry employs over 100,000 persons. In 1989, the industry turned out Rmb 4 billion worth of products and earned $165 million.

In 1989, the four major computer manufacturers - Changcheng, Changjiang, Langchao and Zijin - produced a total of 34,000 micro-computers, representing 57 per cent of total production. The domestic demand for micro-computers is growing and the sales forecast for 1991 shows a significant increase in 1991 (see Table IV.20).

Table IV.20. Microcomputer sales forecast, 1990 and 1991

	1990	1991
Microcomputer installed base	515,000	633,000
Sales volume	105,000	118,000

Source: Ministry of Machinery and Electronics Industries.

Computers have been widely used in the petroleum, chemical, metallurgical, power, textiles, shipbuilding, astronautics and nuclear industries as well as in geological prospecting and meteorological work. China has used micro-computers to turn more than 23,000 ordinary lathes, milling machines, planning and shaping machines and grinders into digit controlled machines during 1985-1989.

The proliferation of computer technology is illustrated by the case of the Capital Iron and Steel Company. Beginning to use computers in 1979, the Capital Iron and Steel Company in Beijing has installed more than 300 computers for production management, accounting, planning and marketing and other office work as well as for production in 17 major mines and factories. Using computers in its automatic system, the ore sintering shop has cut down coke consumption by 12 kilograms for every ton of pig iron produced.

Computers are not only used in scientific research and engineering calculations, but also in information processing, business management, file finding, and industrial and population censuses. The Ministry of Railways, the Ministry of Metallurgical Industry and the Ministry of Astronautics Industry have all established their computer networks as an effective means of modern management.

Electric appliances: rapid diversification

With a large variety of products of different specifications and series, the spectrum of China's electric appliances industry continues to widen. From a dozen varieties of electric appliances in 1978, the number of types of products rose to over 200 in 1989. The production of refrigerators currently encompasses a wider variety, meeting over 30 specifications. The new generation of big and luxurious refrigerators are also produced with additional facilities such as double doors, drawers, a big freezing chamber and automatic display of chamber temperature. In 1978, there were only ordinary desk fans and ceiling fans. Currently, over ten varieties of fans meeting 12 specifications are being produced in China. In recent years, China has also developed quite a number of household appliances such as big-wheel washing machines, automatic washing machines, cylinder-type washing machines, auto temperature regulation rice cookers, electric disk cookers, cooking pans, steamed electric irons, electric vibrators, electric thermal massagers, automatic constant temperature electric blankets, etc. Rapid progress in the production and technology of electric appliances almost eliminated the country's dependence on imports by the end of the 1980s and the industry entered a new era of export-drive.

Exports of electronics and electric appliances: changing export profile

Export earnings from electronics rose significantly from $96 million in 1978 to $2,043 million in 1987, representing a twenty-fold increase - albeit from a low base. During 1985-1987, China's electronic exports doubled each year due to favourable internal factors and the external situation. The revaluation of the Japanese yen and the devaluation of the yuan helped improve the competitiveness of Chinese electronic products.

With an increase in the share of electronics in China's manufactured exports from 1.9 per cent in 1978 to 7.8 per cent in 1987, there had been a significant shift in the composition of electronic exports (see Table IV.21). The changing export profile of electronics was characterized by a fall in the share of electronic components and a significant rise in electronic products, with telecommunication equipment recording a strikingly high rate of expansion. The main source of electronics export expansion stemmed from a significant increase in consumer electronics. The relatively slow growth of components exports in the 1980s was partly attributed to increased domestic demand for electronic components. However, it was largely due to a decline in the importance of defence aid programmes which comprised a large proportion of electronic parts. The increase in the absolute value of components export was due to a sizeable proportion of components and accessories in the export of telecommunications equipment. In 1989 direct export of electronic appliances earned $197 million (see Table IV.22).

Table IV.21. Composition of electronic exports, 1979-1987
(In million $)

	1979	1980	1981	1982	1983	1984	1985	1986	1987
Telecommunications equipment	10.0	16.7	19.5	15.1	35.9	47.3	59.3	145.1	299.5
Consumer products	15.0	42.7	57.1	74.7	104.3	200.5	245.5	506.7	1134.7
of which:									
TV sets	1.1	5.5	6.4	7.2	9.0	10.3	35/7	78.4	201.2
Radios	12.7	31.2	44.4	59.0	79.8	144.9	167.6	333.9	767.4
Gramophones, etc.	1.1	5.7	6.2	8.4	15.5	45.2	42.0	94.2	161.6
Electronic components	97.5	132.6	172.0	151.9	182.1	202.3	186.5	268.6	471.8
Business electronics	8.3	10.6	11.6	16.8	16.3	22.9	36.4	75.8	136.6
Total electronics	130.8	202.6	260.1	258.5	338.5	473.0	527.6	996.1	2042.6

Source: *China: International Trade Annual Statistical Supplement* (various issues).

Table IV.22. Exports of electronic appliances, 1989

Product	Volume (number)	Value ($ million)
Refrigerators	87,200	7.07
Washing machines	104,000	7.49
Air conditioner[a/]	18,300	6.01
Electric fans	10,710,000	124.67
Direct export of electric appliances	..	197.00

Source: *Trade Promotion* (October 31, 1990).

a/ Including parts.

The share of developing countries among the export destinations for China's electronics exports declined from 94 per cent in 1980 to 67 per cent in 1987. Among the developing country markets, the importance of Hong Kong increased and that of Singapore and Malaysia decreased. Electronic exports to Hong Kong rose from 19 per cent in 1984 to 68 per cent in 1987.[22/] The share of developed countries for China's electronics exports also rose. The share of the United States rose from 1 per cent in 1980 to 18.3 per cent in 1987. The so called "Hong Kong effect", i.e., electronics re-exports via Hong Kong, played a significant role in accelerating the export-drive to developed economies. Geographic proximity, non-tariff advantages and Chinese business circles' preference for Hong Kong facilitate the "Hong Kong effect". Hong Kong entrepreneurs have also represented a large proportion of investment in electronics production in Guangdong and Fujian provinces.

Foreign investment: focus on the domestic market

Lured by the unlimited domestic market, attractive incentives and cheap labour transnational corporations have formed a number of joint ventures in a range of labour-intensive electronics production activities. China is using market access as a means to put leverage on foreign firms to import technology. IBM, Hewlett Packard, Digital Equipment Corporation, Wang, Hitachi, Fujitsu, Siemens, Olivetti, Unisys and NEC are among other foreign firms which have directed investment into electronics production in China. In 1990, at least six joint ventures with big transnational corporations were formed. IBM and its joint venture partner, the Tianjin Zhonghian Computer Corporation, are planning to manufacture one of the United States' next generation of personal computers. Sales are targeted largely for the domestic market. Joint ventures in electronics are also aimed at capturing the market vacated by Japan and Newly Industrializing Economies which are rapidly shifting to skill-intensive products.

A good example of direct foreign investment in electronics is the China Zenhua Electronics Industry Corporation which has pursued a policy to occupy the leading position in the country's electronics industry. Headquartered in a mountainous area of Guizhou Province in southern China, this corporation set up a branch in the Shenzhen Special Economic Zone in Guangdong Province in 1981. Since 1981, the corporation has set up more than 30 joint ventures and solely foreign-funded enterprises in the zone. In 1985, the corporation was selected by the State as one of the 10 enterprises in the country to produce programme controlled exchange systems. It set out to find the foreign partner with the most advanced technology in this field and signed a contract with Philips of the Netherlands in December 1987. The Philips integrated business communication system is considered among the most advanced in the world. It is suitable for both large and small enterprises as well as government departments. With a total investment of more than $5 million, the project was expected to go into operation in August 1989. The designed annual production capacity of the system is 150,000 lines. But only 85,000 lines - valued at Rmb 150 million with about Rmb 26 million in net profits - will be set up in the first stage. The company is negotiating with an American company for co-operation in making electronics chips in Shenzhen. Investment for the first stage of the project is likely to be around Rmb 3 million, and 20 million chips were planned to be produced in 1990. Production will be increased to 50 million chips by 1995, 70 per cent of which will be exported. The corporation has linked up with the Second Automobile Works in Hubei Province, one of the biggest automobile manufacturers in China, and the China Huaneng Engineering and Technology Development Corporation in Beijing to set up a joint venture - the Shenzhen Huashen Automobile Electronics Company Limited - and in order to catch up with the rest of the world in the field of advanced technology.

Despite a significant increase in foreign investment flows into the electronics industry, technological sophistication is lagging behind the international frontier. China's technological capabilities continue to be in satellite, precision trajectory measurement, telemetering and telecontrol technologies that were developed on the basis of capabilities acquired through defence electronic applications. Most foreign firms endeavour to accommodate China's increasing demand for transfer of technology. Given the speed with which technology changes, China realizes the need to enhance the technological level and endeavours to create a conducive research environment across several segments of the industry.

The Chinese Government has mapped out a plan to upgrade information and automation technologies with the aid of computer science. China's hi-tech research and development plan, which aims to ensure the country's sustained economic and scientific development in this century and the next, commenced in 1987. To be completed in 15 years, the plan was drawn up by several hundred experts under the guidance of the Leading Group for Science and Technology under the State Council. It covers seven areas: biotechnology, information, space, automation and laser technology and technology for energy development and development of new materials. Each area includes several projects. For instance, information technology covers intelligence computing systems, photo-electric elements, integrated technology for photo-electric systems and information collection and processing technology. Space technology covers large-sized carrier rockets and space technology research and development for peaceful use. Laser technology covers high-property and quality laser devices and application of laser technology in processing and production. Automation technology covers automatic computer manufacturing systems and intelligent robots. Energy development technology includes coal-burning and magneto-hydro-dynamic generation technology, and technology for advanced reactors. Technology for new materials development covers high property structural materials and special function materials.

The Leading Group for Science and Technology has established seven expert committees and completed the organization work necessary to carry out the plan. The biology, energy development and new materials committees have carried out feasibility studies and detailed research plans for projects. Work on major research projects in each area began in 1988.

To perfect the management system for implementing the plan, regulations concerning the management of funds for the hi-tech research plan have been promulgated. The funds will be managed by the State, while a bidding system will be introduced in the management of funds for research projects with funds made available to those units or individuals undertaking research projects. Science funding will be applied to basic research projects.

Prospects: exploring export avenues

China remains a relatively small electronics producer by international standards, accounting for about 0.2 per cent of global electronics production. Foreign joint ventures have proliferated in catering to domestic needs. However, foreign firms seem to be sceptical about the ability of Chinese products to establish a "niche" in the world market. Although China's cheap labour costs remain an attraction, low unit labour cost is not the major determinant of competitiveness. The experience of NIEs suggests that product choice, quality and sequencing, industry structure and competition policy, trade orientation, and diffusion of technology are the major determinants of competitiveness in the electronics industry.

In the preceding two decades, world production of electronics has expanded more rapidly than any other segment of industrial production, leading to the arrival of new products and services marked by rapidly changing technologies. As a result the life cycles of products are being rapidly reduced. Increasing product differentiation has intensified global competition which is becoming increasingly fierce.

Distorted global competition in electronics is in the offing in the face of government backing for loss-making enterprises. The Government of France recently decided to inject FFr 4 billion of fresh capital into ailing computer makers. Electronics producers in Europe fear that this package support might distort competition on the world market. For a marginal producer like China whose electronics industry is primarily oriented to the huge domestic market, these developments in the global electronics industry may have minimal implications. However, China will need to keep abreast of emerging trends on the world market, as the country endeavours to emerge as a leading exporter. According to the contours of the Eighth Plan the government plans to establish three computer export bases. China's share of world electronics exports is expected to rise to 2 per cent by the end of this century. Despite the expected increase in electronics exports, China is likely to remain a net importer of electronics in the 1990s.

The electronics industry is likely to benefit the most from China's fresh offers of deep tax cuts in the network of industrial parks. China is establishing a country-wide network of industrial parks with a view to attracting foreign and domestic investment into high-tech enterprises. In March 1991, the government approved the legal framework for 27 high-tech industrial parks. These parks are aimed at creating a favourable environment for the development of new and high technology in order to facilitate the expansion of high technology industries.

Foreign joint ventures are completely exempt from income taxes for the first two profit-making years. The standard income tax rate for joint ventures is 33 per cent, but high-tech ventures in the industrial parks are subject to a tax rate of 15 per cent. The tax rate will be reduced to 10 per cent for companies exporting over 70 per cent of their total production. There are around 2,500 high-tech enterprises operating in these parks. These enterprises turned out $1.3 billion worth of products in 1990. China endeavours to boost the production of high-tech products of industrial parks from Rmb 7 billion in 1990 to Rmb 25 billion in 1996. These parks are expected to play a significant role in bridging the technological gap in microelectronics and electronic information, aerospace, optical electronics, bioengineering, materials science, ecological science and environmental protection, earth science, marine engineering, medical science, biomedical engineering and energy technology.

Given the critical role of electronics in enhancing the economy-wide competitiveness through the spreading of its benefits across various sectors, the major source of growth will continue to be fuelled by a significant increase in domestic demand. However, a major challenge facing the Chinese electronics industry is to become more competitive internationally. China could endeavour to seize opportunities stemming from global sourcing of components by major players of the global electronics industry who constantly climb up the ladder of high value added products.

I. FUEL AND POWER: ALL-ROUND EXPLORATION AND DEVELOPMENT

The resource base

China is well endowed with fuel and power resources. Recoverable coal and lignite reserves have been estimated at 954 billion tons. Exploration over the past four decades indicates that China has coal deposits of 4,490 billion tons. Coal accounts for 70 per cent of China's total energy consumption. Total oil reserves are estimated at about 5.5 billion tons, making China the fifth largest oil producer in the world. Oil accounts for 21 per cent of total energy production and major oil fields include Daging, Dagang, Shengli, Jizbang, Liaohe, Jiansu, Zhongyung and Jianghan. In addition, some 400 square kilometres in the Tarem, Qaidan and Eichuan basins are thought to be rich in oil and gas reserves. Oil has been discovered in the South China Sea and the East China Sea continental shelf. The Pearl River Mouth is said to be particularly rich in both oil and gas reserves. Substantial offshore reserves occur near the Bohai Sea, the Beiby Gulf and the Ynigge Sea. Increasingly important oil deposit discoveries have been made in the western provinces of Xiangjing and Quinghan. Around 75 per cent of China's crude oil production comes from three onshore producing complexes in north-east China: Danging, Shenghi and Liaohe. China endeavours to speed up the development of nine offshore oil fields in south China and Bohai areas during 1991-1995. Recoverable reserves of non-associated gas have been estimated at 3 trillion cubic feet (tcf), about 90 per cent of which is located in Sichuan province.

China's hydropower potential, among the largest in the world, is estimated at 1,900 TWL per year. Until the late 1980s, about 87 TSL per annum had been developed. China's uranium reserves are sufficient to sustain 15,000 MW of nuclear power plant for three decades. The country is also endowed with abundant geothermal energy in the mountainous south-western region. China is the fourth largest producer and the third largest consumer of commercial energy.

Emerging trends

Coal

China's coal production has grown significantly during the past four decades. With 1,090 million tons of coal in 1990, compared with 600 million tons in 1979, China has become the world's largest coal producer. Concomitant with rising production trends, China's coal productivity increased rapidly during the Seventh Plan period. The daily coal output of a worker rose from 0.939 ton in 1986 to 1.195 tons in 1990, representing an average annual increase of 5.4 per cent. This significant increase in labour productivity was due largely to a 20 per cent increase in the mechanization of coal exploitation.

The share of coal in China's total energy consumption continues to increase, and is expected to reach 76 per cent by the year 2000. Coal production is expected to hit 1.4 billion tons in 2000. Currently, one million tons of coal is turned out by 26 mines in which almost the entire process of coal extraction is mechanized. In order to reduce the country's dependence for coal on the north-west area which has 65.6 per cent of China's coal deposits, there is a tendency to shift coal exploitation to the west. The Ordos coalfields in western China has rich coal deposits. Since the early 1980s, many small local mines have gone into production, and at present about half of the country's coal output is from small mines run by townships, villages and even individuals. With 177.3 million tons of coal exports in 1990, China has emerged as the seventh largest coal exporter in the world. However, only a small proportion of the country's coal production is currently exported.

Oil and gas

China has a total of 18 oil and gas fields. The Daging, Shenghi and Liaohe oilfields each yield between 10 and 5 million tons annually. Sichuan province produces 6 billion cubic metres of natural gas a year. In the 1980s, China set up four offshore oil development companies in the Bohai, South China and Yellow Seas. These signed contracts with 12 countries for offshore exploration covering an area of 150,000 square kilometres. Annex Table A-21 presents oil and gas production data across different areas for the year 1990. During the first seven months of 1990, the production of oil and gas in each area suggests fulfilment of planned targets for the year 1990.

In 1990, China's crude production registered a marginal increase of 0.2 per cent over the same period in 1989. The supply of petroleum products failed to keep pace with demand in the 1980s. As a result, China had to reverse its long-standing policy against imports. With subdued growth of domestic oil production, crude oil supplies have tightened, and since 1988 imports have soared to 110,000 barrels a day in 1989. Exports of crude oil and petroleum products have declined sharply since 1988 (see Table IV.23).

High rates of exploitation in the 1970s and inadequate discoveries are cited as the two main reasons for the declining reserves base. Efforts to achieve oil self-sufficiency are constrained by domestically subsidized oil prices, increasing production costs, excessive oil exports and inadequate spending on exploration. Until 1986, China exported over two-thirds of its increased oil production each year, squeezing the domestic market in favour of easy export earnings. Since then, greater volumes of oil have been allocated to domestic refiners.

The demand for petroleum products is expected to grow between 65-100 per cent by the year 2000, implying 3.5-4.5 million bpd.[23/] In contrast, crude production is expected to rise by only 2.3-3.5 million bpd. Unless efforts to boost exploration and development are intensified, China will need to depend on imports in order to make up this shortfall. China is intensifying efforts to expand

Table IV.23. Oil production and exports, 1975-1990

Year	Crude production (Million b/d)	Exports: Crude oil (1,000 b/d)	Exports: Oil products (1,000 b/d)
1975	1.541	198	44
1976	1.743	170	39
1977	1.873	182	39
1978	2.081	226	43
1979	2.123	269	61
1980	2.119	266	84
1981	2.024	277	92
1982	2.042	304	105
1983	2.121	298	108
1984	2.296	446	120
1985	2.498	601	124
1986	2.622	570	109
1987	2.683	545	99
1988	2.740	550	94
1989	2.750	490	90
1990	2.756[a/]	391	..

Source: *Oil and Gas Journal* (January 14, 1991).

a/ From Central Intelligence Agency, *International Energy Statistical Review*.

Fig.IV.9. Production of crude oil, 1980 - 1990 (Million barrels per day)

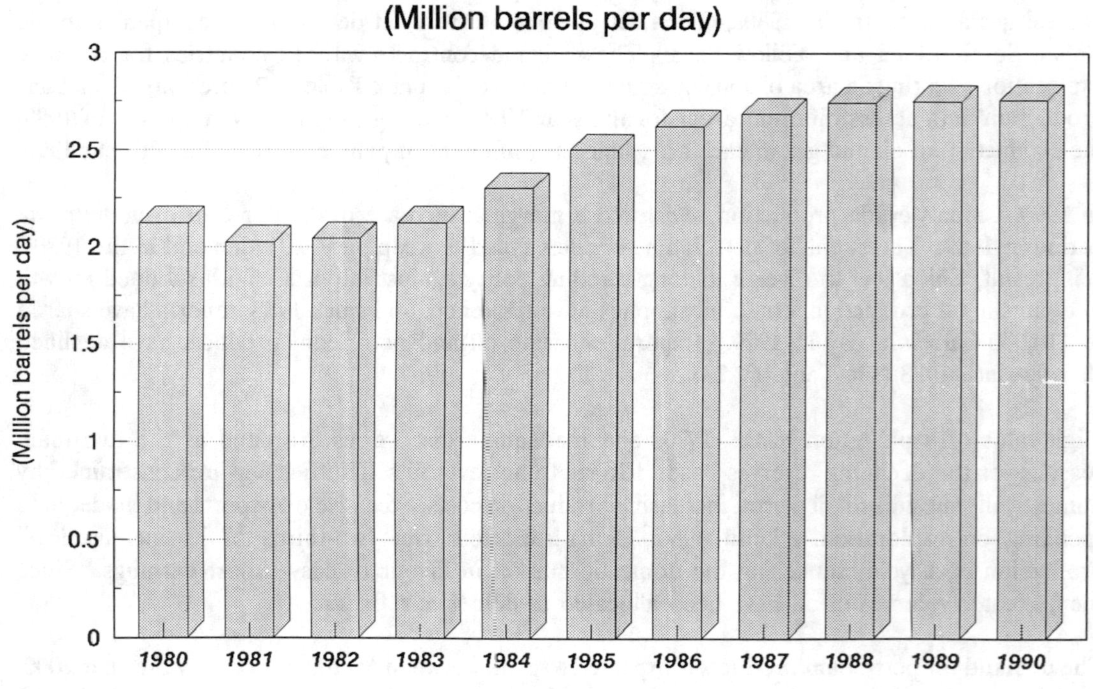

the reserves base in order to enhance the production of crude oil by more than one-third of its present levels. Indications are that the government has planned to spend $20 billion for oil exploration and development during 1991-1995.

There are numerous reports of new oil strikes. However, new production is hardly keeping pace with rising demand. China's ambitious target of producing 4 million bpd of oil by the year 2000 hinges on the level of foreign investors' participation in oil exploration and development. China's controlled crude price is about 21 per cent of the international price. This is too low to recoup costs and generate cash flow for exploration and development efforts. The country's offshore region is an area of strong interest for foreign investors. Oil agencies plan to accelerate activities in nine oilfields in South China and Bohai Seas during the Eighth Plan. China Oil and Natural Gas Exploration and Development Corporation (CONGEDC) is aiming at production targets of 100,000 bpd of crude oil and 116 million metres of gas from offshore fields by the end of 1992.

Power[24/]

China has around 100 large power stations, each with an installed capacity of over 250,000 kW. The Kengkou Thermal Power Plant in Shanxi was built to take advantage of the rich coal fields in the province. The Czechoulia Hydroelectric Power Station on Changziang has an installed capacity of over 2.7 million kW and is the largest of its kind in China. The south-western part of the country is using foreign funds to develop its rich water resources. The Lubuge Hydroelectric Power Station in Yunnan province received foreign investment to the tune of $1.9 billion by 1987. Nine other stations with an estimated capacity of 12 million kW have also attracted foreign investors.

In the sphere of exploring other sources of energy, China is developing nuclear power, with an accent on safety. Following the completion of China's first nuclear power station at Quishan in Haiyan in Zhejiang province in 1990 with a capacity of 1.5 billion kWh a year, four more nuclear reactors with installed capacity of 2.4 million kW are planned to be installed in Quishan.

Taking advantage of the water power of the Huangha (Yellow River), China has been carrying out a plan to construct 15 hydropower stations at the middle-upper reaches of the river. Electricity generation is inadequate to meet needs, a major cause of underutilization of capacity in Chinese manufacturing.

The government is well aware of the need to enhance fuel production and power generation in order to achieve the aim of doubling the country's GNP by the year 2000. To achieve this target, the country will need 1.4 billion tons of coal, 200 million tons of petroleum and 1,200 billion kW of electricity by the turn of the century. China is determined to meet this challenge. A number of large- and medium-sized power stations will be constructed, among them more hydroelectric stations on the Yellow, Yangtze and Hongshin rivers. Thermal power stations will be located in coastal areas and along coal-transporting railroads. Nuclear power stations are planned to be located in areas short of other sources of energy. Large coal mines will open up in Inner Mongolia, Shianzi and Shanzi, while oil exploration will continue in Bohai Bay and other offshore areas and in the Tarmi Basin. The plan for the 1990s is to explore and develop new sources and to achieve efficiency in energy use.

Energy consumption and efficiency

China's total energy consumption ranks third in the world's energy use after the United States and USSR. Currently China accounts for around 8 per cent of the world's energy use. China's consumption of energy per unit of GDP is quite high, though per capita consumption is low. The energy intensity of industrial operations depends upon the efficiency of fuel use, the structure of industrial production, the scale of industrial units, raw materials used and the technology employed. In the 1980s, the government began to address the issue of inefficient energy use through industrial modernization, technical assistance and conservation centres. Energy consumption per unit of gross value of industrial output fell at an average annual rate of 7 per cent in 1979-1981, 3 per cent in 1982-1983 and 7 per cent in 1984. Around 40 per cent of these energy savings stemmed from incidence of technical progress. However, about 60 per cent of energy savings was attributed to structural changes in terms of a decline in the importance of heavy industry and the closure of inefficient small-scale plants.

An analysis of overall industrial energy intensity is constrained by paucity of pertinent data. The energy efficiency of the Chinese steel industry may be taken as a representative case of energy intensity in heavy industry. A recent study[25/] compares the Chinese iron and steel industry of 1987 in terms of energy consumption and production data with the United States industry to highlight the major differences. The findings of the study merit attention. The ratio of total energy consumed by the Chinese iron and steel industry to production of steel products reveals an energy intensity of 2 tons coal equivalent per ton. The adjusted average energy intensity[26/] of the Chinese steel industry in 1987 of about 1,450 kg per ton of steel mill products is about 35 per cent higher than that of the comparable United States industry. In China a large number of small plants which use antiquated technology account for only 25 per cent of steel production. The adjusted energy intensity of the more modern key plants is 1,280 kg coal equivalent per ton; about 20 per cent more than that of the United States. Clearly, it is important for China to take advantage of new avenues which have been developed in other countries in order to enhance efficiency in the use of energy and material.

J. RURAL INDUSTRY: A DYNAMIC FORCE IN INDUSTRIAL TRANSFORMATION

An overview

The ideological foundations of the rural economy of China have changed from central planning and large agricultural collective farms to market mechanisms and independent farms and firms. Rapid growth of rural industrial and other non-agricultural activities has been accompanied by significant increases in rural personal incomes and major shifts in the structure of the labour force. Fresh employment opportunities in non-agricultural activities[27/] absorbed over 30 million persons during 1978-1986. During the same period, gross value of rural industrial output grew by 23 per cent per annum in real terms. Among rural enterprises, the township- and village-owned enterprises and farmers' joint-stock enterprises assume a dominant position.

The degree of rural industrialization across the Chinese provinces in the mid-1980s can be gauged from the share of rural industry in rural social value of output (RSVO), i.e., the gross product of rural society[28/], and from rural employment in each province. Annex Table A-22 shows that Beijing, Zhejiang, Jiangsu, Tianjin and Shanghai were the most industrialized regions in terms of rural industries' contribution to the total value of production in the rural economy. In this most industrialized group of regions, over 50 per cent of RSVO originated from rural industry. The share of rural industry in rural employment in these regions ranged from 45.8 per cent in Shanghai to 19.6 per cent in Zhejiang and Jiangsu. In relative terms these indicators reveal a higher degree of rural industrialization in these regions. Annex Table A-22 shows the provincial contrast in rural industrialization; nine provinces are least industrialized both in respect of the value of rural industries' output and employment.

Despite a rather unbalanced regional development of rural industries, the emergence of millions of rural firms in the preceding decade contributed a new force in reshaping the structure of the Chinese rural economy; new activities and technologies have become operative along with the emergence of a large number of entrepreneurs, managers, technicians and skilled labour force. As a result, rural industry has been the most rapidly growing segment of Chinese industry. These industries also became an integral component of the rural economy in the 1980s.

In 1989, rural industries produced 348 million tons of coal, 65.3 million tons of cement, 4.9 million tons of machine-made paper, 745 million sets of farming tools, 2.33 billion suits of clothes and 13.07 billion kWh of electricity. During 1979-1989, rural enterprises ploughed more than Rmb 80 billion from their profits into rural development projects for education, health care and welfare facilities. Recently, over Rmb 500 million of rural enterprises' profits has been used annually for town construction. By 1989 rural towns numbered 12,000 compared with 2,600 in 1980.

Growth impulses on the supply side stemmed mainly from dramatic changes in rural policies and the institutional framework. On the demand side, rural enterprises were quicker than State enterprises in filling numerous "vacuums" in existing demand. Rising personal incomes coupled with enterprise reforms generated demand for both consumer and investment goods, creating significant opportunities for rural enterprises.

Rural enterprises also constitute an outlet for the surplus labour force. In 1989, 93.67 million persons worked in rural enterprises, representing 23 per cent of the total rural labour force or 62 per cent of the increased labour force between 1978 and 1989, or half of the current total labour force. It was estimated that by 1990 rural enterprises would employ 100 million workers. Having realized the inevitable role of rural enterprises, the government endeavours to support actively the further expansion of these enterprises on sound lines. During the high-speed phase of expansion in the 1980s, rural industries revealed great dynamism to demonstrate their ability to take advantage of opportunities. Having grown significantly in 1988, rural non-agricultural output expanded less rapidly. There are indications of a vibrant growth in the Chinese rural industry in the first four months of 1991. The current readjustments call for a rational industrial structure and product mix, improvement in management, the use of up-to-date technology and economic efficiency.

Performance

The performance of township and village enterprises in terms of output, employment, assets and profits has been spectacular during 1983-1990 (see Table IV.24). The value of gross output in nominal terms grew at an average annual rate of 41.5 per cent during 1983-1990. Given the general assumption about the low rate of inflation until the mid-1980s, the growth of real output must also have been significantly high. Concomitant with the rapid growth of output, employment grew significantly. Total wage bills, fixed assets, profits and tax payments also increased substantially in line with rising output and employment. In relative terms, profits have risen much more slowly than other performance indicators of township- or village-owned community enterprises.

During 1978-1989, Chinese farmers gained one-fourth of their net income from rural enterprises. During 1986-1989, the proportion rose to 50 per cent. The net increase in taxes from rural enterprises during 1984-1989 accounted for 50 per cent of the increase in State financial revenue. During the same period, the total export value of rural enterprises was to the tune of $35.5 billion; $10.5 billion in 1989 alone accounting for about one-fifth of the total export earnings of China. In the face of rapid growth and significant strides achieved by rural industries, the contribution of these industries to the gross product of rural society rose significantly from 31.4 per cent in 1978 to 58 per cent in 1989. Thus rural industries constitute a force to reckon with for the industrial transformation of Chinese economy. According to available statistics, township enterprises have hitherto directly invested Rmb 26.7 billion in agriculture for purchasing machinery, disseminating scientific and technological results, and strengthening infrastructure, contributing significantly to the modernization of agriculture.

Constraints and prospects

After a decade of rapid growth, rural industries currently face increasingly severe competition and market saturation, leading to a number of investment failures, inefficient technology and low quality of products hinder the ability of rural enterprises to compete. Factor immobility instills

Table IV.24. Selected performance indicators of township and village enterprises, 1983-1990
(Percentage)

	1983	1984	1985	1986	1987	1988	1989	1990
Output								
Rural non-agricultural output	19.20	68.16	59.57	29.78	33.95	36.95	14.36	13.91
Industrial output value	17.19	64.49	46.72	32.08	34.40	39.65	15.78	15.37
Number of firms								
Township-village-based community enterprises (TVCEs)	-1.15	22.52	-4.88	-3.29	4.04	0.72	-3.42	-5.32
Industrial TVCEs	-0.68	21.12	-5.15	2.68	10.45	2.90	-1.43	-4.78
Employment								
TVCEs	3.91	18.97	7.90	5.77	7.08	4.07	-3.55	-2.71
Industrial TVCEs	4.60	17.56	9.15	9.32	9.79	0.34	3.03	-1.47
Financial indicators[a/]								
Gross revenues	20.61	36.55	44.10	21.68	31.96	44.24	13.93	8.23
Taxes	30.84	34.39	37.22	26.86	22.03	40.71	15.22	1.08
Profits	1.57	9.25	33.14	-6.04	16.59	38.05	-7.36	-3.08
Wage bill	14.91	36.11	25.97	17.94	20.30	26.54	7.30	4.49
Fixed assets (year-end)	10.87	20.89	31.02	25.67	29.56	29.16	21.23	14.65
Bank loans	19.15	102.62	40.33	46.96	41.37	26.90	18.14	22.06

Source: State Statistical Bureau of China, *Statistical Yearbook of China*; and *China Rural Statistical Yearbook* (various issues).

a/ For TVCEs only.

Fig.IV.10. Selected performance indicators of township and village enterprises, 1983 - 1990 (Annual percentage change)

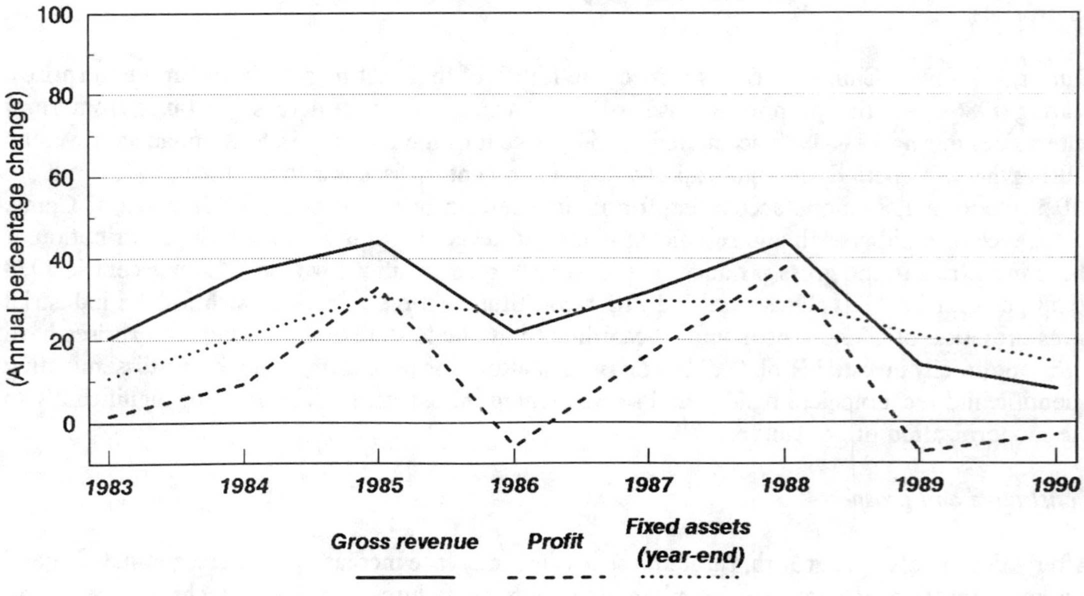

a source of rigidity, inhibiting the diffusion and adaptation of advanced technologies and practices. An inadequate network of support services is one of the constraints inhibiting the further growth of even successful enterprises.

The average size of rural firms has not increased over the years. Although economies of scale do not necessarily constitute the determinant of efficiency in the case of many small-scale industries, in the production of products such as cement and steel small size is generally a barrier to efficiency gains.

As to the technological constraints, the problems differ between industrially developed regions and least developed regions. In developed regions the existing technological base is stronger. In the industrially less developed regions, there is a need to augment human skills to choose, adapt and to manage technology. The problem is not related to acquiring technology, but to choosing and adapting technology appropriate to local conditions. The implementation of the "Spark Plan" in the country's rural areas has promoted a number of local resource-based key industries. The "Spark Plan" is aimed at the gradual transformation of traditional agricultural patterns of scattered management into an intensive and operational type of farming, poultry farming, forestry and fruit growing in order to form commodity production centres. The "Spark Plan" particularly emphasizes the establishment of rural industries based on agricultural and mineral resources. The purpose is to combine farming and breeding with the processing of agricultural and animal products, and mining with the processing of mineral products. The plan has injected new technologies into the production process of rural enterprises, helping them raise the quality of products, reduce energy consumption and enhance the rational use of raw materials. Attention is also focused on preventing pollution in order to protect local ecology.

Between 1986 and 1989, over 20,000 projects were launched throughout the country within the framework of the "Spark Plan". Of these, 10,346 projects have been completed. The resultant newly increased output value was to the tune of Rmb 22.18 billion in addition to Rmb 5.53 billion in profits and taxes. Exports from these projects fetched around $740 million in 1989, equivalent to 45 per cent of the total export volume of township enterprises in the preceding three years.[29/] In early 1991, around 57,800 rural enterprises earned foreign exchange to the value of Rmb 9.5 billion. Thus, the implementation of the "Spark Plan" in rural China ignites a fresh stimulus with advanced technologies yielding quick results. Given the massive government support for rural industries to evolve and adapt to the changing trends, the long-term prospects for rural industries are bright. The government urges rural industries to look for possibilities of greater co-operation with their counterparts overseas.

K. THE ROLE OF TECHNICAL CO-OPERATION

The total net official development assistance (ODA) channelled to China rose from $669 million in 1983 to $2,227 million in 1989. External assistance received by China in 1989 consisted of investment project assistance ($1,506 million), free-standing technical co-operation ($216.8 million), food aid ($81.9 million), investment-related technical co-operation ($53.4 million), and emergency and relief assistance ($24.4 million). As of end-1990, UNIDO has developed and approved about 197 projects in China with a total value amounting to approximately $60 million under various financing resources. Of this, 113 projects were operationally completed and 65 projects are under implementation.

Multilateral assistance fell by 6.8 per cent in 1989, while bilateral assistance rose by 24.8 per cent in the same year. Japan accounted for 59 per cent of bilateral assistance in 1989, followed by France (14.0 per cent), Federal Republic of Germany (6.5 per cent), Italy (4.1 per cent), and Austria (4.1 per cent). Other bilateral donors accounted for 12 per cent of bilateral assistance.

China's economic and social development in the coming decade, especially during the Eighth Five-Year Plan period (1991-1995) will focus on: strengthening agriculture, enhancing basic industries and infrastructure; readjusting and transforming processing industries; and strengthening education, science and technology. In line with these objectives and the principles stated above, five areas of concentration have been identified for the Country Programme.

- Rural economic development and production;
- Energy, transport and communications;
- Industrial productivity and profitability;
- Social development and quality of life; and
- Economic and public administration reform and management.

These areas are closely inter-related, and will together contribute to the achievement of China's economic and social objectives in the coming decade, and especially during the Eighth Five-Year Plan period.

Individual programmes to be assisted by UNDP will be selected from within the five areas of concentration stated above and will be designed in a programme approach. It is recognized that in order to maintain flexibility, not all UNDP resources can be channelled through programmes. The interventions proposed within each of the programme area identified, therefore, are not intended to be exhaustive, but will constitute by far the bulk of the programme.

Five key multi-sectoral themes will permeate the areas of concentration mentioned above. These are:

- Management and administrative performance;
- The environment;
- Development of poor areas and minority nationalities;
- Increasing the role of women in development; and
- Technical co-operation among developing countries.

Just as it is necessary to see the inter-relationship among the areas of concentration, it is equally important to see the relationship between areas of concentration and the multi-sectoral themes. These themes will receive constant attention in the numerous interventions that will be made in each of the areas of concentration. The themes themselves are mutually supportive and are bound by the common need to increase efficiency in the use of human, natural and financial resources. The proper application of fertilizers, for example, will increase yields, reduce pollution, save money, reduce waste, improve investment planning in fertilizer production plants and reduce imports.

Management

Planning, administration and management of sectors, subsectors, individual enterprises and institutions will be strengthened in the overall context of the Country programme. This theme is established in the area of concentration of economic and public administration reform and is the principal objective of five other programmes dealing with forestry, water, rural and large- and medium-sized enterprises and health.

Environment

Sound environmental policy is more than pollution control. It involves development planning which preserves the environment through improved economic efficiency and sound economic and investment decisions. The new Country Programme will provide assistance in (1) addressing global issues such as the use of ozone depleting substances; (ii) the implementation of national environmental strategy; and (iii) new projects in all areas of concentration. The environment, as an integral part of development planning permeates the entire Country Programme. This is especially evident in problems associated with water, fertilizers and pesticides, coal utilization, energy conservation and environmental health.

Poor and remote areas

UNDP assistance will help augment rural economic production and the quality of life in poor areas, especially whose inhabited by minority groups. To the extent possible, these efforts will be made in collaboration with international non-governmental organizations. The areas of poverty concentration will be prime beneficiaries in the programmes targeted on food production, especially in the arid and semi-arid zones, training of managers and officials in poor counties and the social development programmes in health and education.

Women in development

Women will continue to participate fully in and benefit from the UNDP programmes. The aim will be to further integrate the concerns of women within the overall programme rather than isolating them in a separate set of projects. The two programmes that best lend themselves to the incorporation of specific components aimed at women are rural economic production and social development. More importantly, care will be taken when elaborating these and other programmes to ensure that where appropriate the concerns of women are addressed at the design stage.

Technical co-operation among developing countries

China has always attached great importance to technical co-operation among developing countries (TCDC) and has made tremendous efforts to promote it. Considerable scope exists to fully incorporate TCDC as a mechanism for gaining access to intermediate technology within the Country Programme. China will continue to advance TCDC, and at the same time to seek more extensive participation in the regional, interregional and global programmes.

As long as China's population continues to grow at the rate of 17 million per annum, and the fact that 80 per cent of the population live in rural areas, rural economic production will be fundamental in the country's economic growth. Critical to this growth are food production, rural enterprise management, forestry management, and water resources planning and management. The programmes to be developed in this area of concentration will advance solutions to these broad problems, as well as those relating to the multi-sectoral themes on management, the environment, poverty alleviation in poor and minority areas and the role of women in development.

The poverty and environmental aspects of this area of concentration make it an obvious focus of attention by other donors, and it is expected that UNDP will work with bilateral donors and the multilateral grant and loan institutions. Co-operations with the latter will be in the form of pre-investment project preparation work and technical assistance in direct support of investments.

Twenty programmes for technical co-operation during 1991-1995

Food production

This programme is expected to enhance the sustainability of agriculture through more effective management and the efficient use of environmental resources. It will also increase food production and incomes in a number of areas in the country including both poor and relatively advanced ones while reducing desertification. It will channel assistance to activities of particular benefit to women, and will include opportunities for technical co-operation with developing countries which are particularly advanced in arid and semi-arid land agriculture.

Forestry management

An effective national forest resources management and monitoring programme will be developed by the setting up of a model demonstration computerized systems for data collection, compilation, analysis, and retrieval, as well as human resource development. Assistance will also be sought from other multilateral and bilateral sources within the framework of the programme.

Water resource planning and management

Water resource allocation and management in northern China through the introduction of mathematical planning models at the central level and sub-models at the provincial and municipality levels will result in improved decision making at the national and local government levels. As a part of this programme, specific studies will also be carried out with respect to water conservation for agriculture, ground water pollution and waste water treatment.

Rural enterprise and development management

In order to increase the pace of development in poor areas of the country, it is essential that provincial and county level officials be trained in basic aspects of development economics, economic planning, programme identification, formulation, appraisal (including cost benefit analysis), evaluation, and other skills essential for the management of the process of development. Officials will also be trained in essential skills such as alternative methods of environmental impact assessment for the integration of environmental concerns in the development process with a view to fostering sustainable development. Particular attention will be paid to ensuring that women officials will constitute an equitable proportion of those trained.

Coal resource utilization

The programme will result in improved planning, allocation and distribution of coal nation-wide; improved occupational safety at the site of coal extraction; pollution reduction and improved efficiency through the introduction of techniques such as washing and gradation at the pit site and improved coal combustion technologies. This programme is expected to include aspects of pre-investment project preparation and should attract cost-sharing contributions from other multilateral and bilateral sources because of its environmental nature and donor interest in providing relevant technology.

Energy efficiency and conservation

The programme will focus on policy research and management aspects of the sector, particularly the development of medium- and long-term plans for energy production and utilization; the adoption of rational pricing policies; energy transportation; labour productivity and economic efficiency in energy production plants; and the development and application of laws and regulations relating to the energy sector.

The overall impact of the programme will be to increase the efficiency of energy utilization in targeted industries by up to 40 per cent, and broaden the range of energy sources available as clean alternatives to coal. The programme will reduce waste in oil extraction in certain relevant declining fields by at least 20 per cent. Multi-sectoral themes of the Country Programme addressed by this programme are management and the environment, with impact on poor and remote areas through increased use of renewable sources of energy already introduced in the Second Country Programme.

Telecommunication modernization

The programme, with catalytic inputs from the IPF, will aim at rationalizing the use of different communications technologies; the introduction, adaptation and development of new technologies such as optical fibres, integrated digital networks and satellites; and the management of such communication systems. The programme will serve to raise the efficiency and reliability of telecommunications facilities through the introduction of new technologies and the strengthening of management capabilities. It will also result both in better access to communication facilities for remoter parts of the country, and the availability of a broader range of services to the user.

Transport design

This relatively small but important programme will primarily address the technological aspects of the sector, particularly the technical design, development and testing of different modes of transportation including rail, road and shipping as part of a larger ;programme requiring capital investment, including public sources and foreign direct investment. The programme will result in the development of new and improved mass transit bus and car designs. Improvements will also be made in the design of locomotives and ships used for the transportation of coal and other primary products. In addition, the programme is expected to contribute to the elimination of transportation bottle-necks in basic industry and eventually to rapid economic expansion.

Large- and medium-sized enterprise management

Enterprise reform and improved enterprise management are essential for the sector and for the economy as a whole. A network of training centres will be established to serve the poorer provinces of the Northwest. The network will focus its activities on developing broad management education curricula as well as more specialized, short-term courses for senior managers of large- and medium-sized enterprises situated in the area. The programme will endow enterprise managers with a sound understanding of approaches and techniques that can be applied to increase efficiency and profitability, including environmental aspects. Women will constitute a significant proportion of all management trainers as well as trainees.

Machine tool industry

The machine tool industry has not kept pace with technological developments in the rest of the world over the last two decades, and relies mainly on hand operated equipment. This programme will improve the quality of a range of essential machine tools, including computer numerical control systems. The programme will also introduce new management techniques aimed at minimizing costs while improving overall quality standards. Programme impact will ultimately be judged by increased productivity of processing and manufacturing industries which use the improved machine tools.

Industrial materials

The development of materials with particular properties to meet the requirements of different industries is a special constraint in the sector and a high priority for attention. A range of materials serving the needs of the electronics, paper, and dye production industries will be focused upon. The programme will result in the development of new materials, improved techniques for the manufacture of currently produced materials, improved quality of the materials produced and improved linkages between research and development facilities and actual production.

Education systems and administration

The programme will serve to improve the management of scarce resources and the generation of additional resources for increasing the effectiveness of the education system and will cover both the efficient and effective deployment of human as well as financial and physical resources. Education administrators at the school level who are drawn largely from the ranks of teaching faculty will be trained in the necessary management skills to conduct their management and administrative functions more effectively. The management and administrative training requirements of the staff of basic education institutions at the provincial and municipal levels will be assessed and programmes for the training of trainers established, concentrating on those provinces and municipalities with resource constraints which are acknowledged as being particularly severe. Developing an environmental education component of the regular curriculum for basic education will form part of the programme.

Teaching methodologies and materials

Within the framework of education for all programme and the government's nine-year compulsory education programme, the skills and teaching techniques of primary and middle school teachers will be upgraded. The programme will result in switching the curricula for key education programmes to more interactive methods of education and increasing retention on the part of pupils. A core cadre of teacher trainers will be developed with a view to rapidly training a large numbers of teachers in the application of the new teaching curricula and teaching methods. The programme will concentrate assistance on areas which need it most; areas densely populated with minority nationalities.

Health management and delivery systems

Shortage of cadres of intermediate level health personnel such as nurses are a critical bottle-neck for the effective implementation of health care. A sustainable network for the training of nurses at different levels will be established along with the capability to plan and forecast nurse manpower requirements. The programme will also result in changes to the current nursing system including recruitment and policies regarding within-service grade and functions. It is expected that this programme will attract supplementary funding from bilateral donors because of its obvious human resource development orientation.

Critical diseases

Approximately 380 million people live in iodine-scarce areas, and suffer from varying degrees of iodine deficiency resulting in symptoms such as goitre and cretinism. UNDP assistance will be drawn upon to support the establishment of an effective system to assess the patterns of incidence and the prevention and treatment of such diseases. The programme will result in better trained manpower in epidemiological techniques and the treatment of advanced iodine-deficiency. Preventive programmes will also be designed with UNDP assistance.

Environmental protection and health

This programme addresses specific aspects of the environment which have a direct impact on health. Linkages will be made with programmes in other areas of concentration, particularly in the industrial sector. It is to be expected that considerable interest will be shown in this programme by other donors, particularly in the sharing of existing technologies, and the promotion of direct co-operation between Chinese and foreign enterprises.

Employment, social security and vocational training

Full employment poses a challenge to economic efficiency and social stability. To establish a dynamic labour market, government's capability to undertake manpower planning on the basis of demand, primarily in the productive sector, will be strengthened. Vocational training facilities will be established in a manner to ensure that they respond to changing needs in the economy and are able to match skills to changing requirements. Under the same programme of assistance, UNDP resources will be allocated to the development of social security systems on a pilot of experimental basis. The capacity of the government institutions to manage and administer social security financing systems will be strengthened and the systems themselves defined and established. This again will constitute an especially important focus of UNDP assistance under the Third Country Programme. Special attention will be given to women to ensure that they are fully eligible for benefits under social security schemes and retraining programmes in the vocational training component.

Public administration reform

Strong economic performance in China assumes a corps of professional and well trained civil servants and a rational set of public institutions with well-defined functions at different levels of government. Reforms in this area will have a profound impact on the country's future. The government is in the process of establishing a new civil service system based on merit through competitive recruitment and performance evaluation. A careful assessment of job functions will lead to revision of job descriptions and redeployment of staff. The new system will apply to all public institutions throughout the country and to the major State-owned enterprises. This will be supported by effective training of a core cadre of civil servants and mid-career and senior cadres through the establishment of the National School of Administration, among other mechanisms. In addition, the structure of government at provincial, municipal and country level will be rationalized also with reference to its inter-relationship with national government. UNDP technical assistance will be selectively provided in these areas.

Economic sector management

Government's efforts to improve the effectiveness of macro-economic management by strengthening economic instruments (such as fiscal and monetary policy), administrative systems, and laws and regulations will continue to be supported. The programme will support the government's decision making in the deepening of economic reform through international exchanges as an input in the review of policy options as well as in the detailed elaboration and implementation of reform measures.

Aspects of city management will be addressed through the practical application of modern urban management approaches. City leaders will be in a better position to address critical issues such as the provision of transportation, communication, health, insurance, utilities and other essential services. The programme will also address the control and treatment of pollution resulting from domestic, industrial, and agricultural activities associated with the cities in question.

Foreign trade and investment management

UNDP assistance will be provided to train a core group of personnel in trade and management issues. Assistance will also be provided to select industries in further developing their export potential. In the area of investment management, select provinces and municipalities on the eastern seaboard will be able to use foreign investment for development more efficiently. Personnel will be trained, and advice will be provided in assessing the types of technologies and management skills that can best be acquired through co-operation with foreign investors.

The government continues to strengthen its co-ordination of external assistance to ensure that it is used to best serve national development priorities. Joint sectoral studies or reviews may be organized by the government in a selective manner to facilitate the elaboration of programmes to be supported by more than one donor. As the requirements for international technical co-operation in support of priority areas addressed by the Third Country Programme exceed UNDP resources, special efforts will be made to secure contributions of other donors within the framework of clearly defined programmes.

In order to strengthen national capacity to execute projects, steps will be taken to assist the government in computerization for more effective management, strengthen its monitoring capacity and replace the direct payment system currently in place with advance payment in order to fully transfer operational responsibility to the government. In addition, a newly developed Project Implementation Manual will be finalized and distributed and serve as the definitive reference for National Project Directors and National Programme Managers. More seminars and workshops will be held to develop the capacity of national counterpart personnel in relevant line ministries and project personnel. Management systems within the government will be further strengthened in order to improve the quality of execution. There will be an increased use of national consultants.

By the end of 1990, around $49 million had been committed mostly for the year 1991, as a part of resources for the Second Country Programme Projects (1991-1995). There is an increased awareness of the importance of sustainable development in the 1990s. The elimination of critical constraints in the energy, transportation, raw materials, basic industry and infrastructure sectors is crucial for sustaining an average annual industrial growth rate of 6.5 per cent envisaged for the period 1991-1995.

NOTES TO CHAPTER IV

1/ For details, see "Nuclear technology boosts agriculture", *Beijing Review* (September 17-23, 1990).

2/ See *China Food Industry Almanac, 1989.*

3/ For details, see *Food processing* (August 1988), pp. 112-114.

4/ *China Daily* (Beijing, 20 January 1988).

5/ See Tai Ming Cheung, "Bale of tears", *Far Eastern Economic Review* (22 February 1990), p. 66.

6/ China imported 900,000 tons of cotton annually in the late-1970s.

7/ For industrial enterprises with independent accounting systems.

8/ Currently man-made fibres constitute 22 per cent of all fibres. See *Textile Horizons* (February 1990), p. 22.

9/ See *Beijing Review* (March 1990), pp. 12-18.

10/ Economic Information Department, *Trade Promotion* (June 16, 1990), p. 6.

11/ In the first two months in 1990 the price of steel sheet was Rmb 3,500 per ton, compared with Rmb 4,000 per ton of imported steel. *Ibid.*, p. 6.

12/ See Guang Xi Jin, "Productivity improvement in Chinese iron and steel industry", *Steel Times International* (May 1988), p. 52.

13/ Detailed information on oil and natural gas as sources of energy in China is presented in Chapter IV.

14/ See World Chemical Industry Yearbook, *China Chemical Industry 1988* (Beijing, February 1988), p. 562.

15/ *Fertilizer International* (January 1990), No. 281.

16/ Information relates only to independent accounting enterprises.

17/ See *China Machinery Industries Yearbook 1989* (Beijing, 1989), p. 39.

18/ The only exception is transport equipment.

19/ *Trade Promotion* (October 31, 1990), p. 13.

20/ Defined as the number of television sets per household.

21/ For details, see Wen Tianshen "China's computer industry on the rise", *China Today*, (February 1, 1991), pp. 40-42.

22/ Calculated from Hong Kong trade statistics.

23/ See David G. Fridley and James P. Darian, "China's resource crisis", *Far Eastern Economic Review* (27 September 1990), p. 62.

24/ Draws largely on information contained in *China Today* (February 1990), pp. 64-66.

25/ Marc Ross and Liu Feng, "The energy efficiency of the steel industry of China", *Energy* (Great Britain, 1991), Vol. 16, No. 5, pp. 833-849.

26/ The adjustment includes the effect of very large production of cast-iron products in China. For details see Marc Ross and Liu Feng, *ibid.*

27/ Industry, construction, transport and commerce.

28/ For details, see Kenneth R. Walker, "Forty Years on: Provincial Contrasts in China's rural economic development", *The China Quarterly* (September 1989) No. 119, pp. 448-474.

29/ *Beijing Review* (July 16-22, 1990), p. 27.

ANNEX A

STATISTICAL TABLES

Annex Table A-1. Recent economic trends, 1985-1990
(Annual percentage change unless otherwise specified)

	1985	1986	1987	1988	1989	1990	Average annual growth rate (1986-1990) (Percentage)
Real GNP growth (Percentage)	12.7	8.3	11.0	10.8	3.9	5.0	7.8
Real industrial growth[a] (Percentage)	21.4	11.7	17.7	20.7	8.5	7.6	13.1
Real agricultural growth[a] (Percentage)	3.4	3.4	5.8	3.9	3.1	7.6	4.7
Fixed asset investment[b] (Percentage)	38.8	18.7	20.5	23.5	8.0	7.6	10.4
Real consumption growth[c] (Percentage)	15.7	7.6	7.2	8.2	0.8	3.8	5.7
Consumer price inflation (Percentage)	8.8	6.0	7.3	18.5	17.8	2.1	10.1
Population (Billion, end year)	1.059	1.074	1.089	1.104	1.112	1.143	1.540
Money supply growth (Percentage)	19.9	27.9	18.5	20.0	6.3	..	..

Sources: State Statistical Bureau of China, *China Statistical Yearbook 1989* (Beijing 1990); IMF, *International Financial Statistics*; press reports.

a/ In gross value terms.
b/ Investment in fixed assets in current Rmb.
c/ Based on indices of consumption in national income, 1952 = 100.

Annex Table A-2. External payments, 1985-1990
(In billion $)

	1985	1986	1987	1988	1989	1990
Merchandise exports[a]	27.4	30.9	39.4	47.5	52.5	62.1
Merchandise imports[a]	42.3	42.9	43.2	55.3	59.1	53.4
Trade balance[a]	-14.9	-12.0	-3.8	-7.8	-6.6	8.7
Invisibles and transfers balance	0.1	-0.2	-1.2	-1.1	-0.02	..
Current account	-11.4	-7.0	3.1	-3.8	-4.3	..
Capital account	6.7	5.9	6.0	7.1	3.7	..
Change in reserves	4.6	1.3	-4.9	-2.2	0.6	..
Exchange rate (Rmb per $, average)	2.9	3.4	3.7	3.7	3.7	4.8
Total external debt	16.7	23.7	35.4	40.0	41.3	52.5
Long term	10.0	16.6	26.0	32.7	37.0	..
Short term	6.4	6.1	8.2	7.3	4.3	..
Debt service ratio (Percentage, long term only)	6.3	8.2	7.5	9.0	8.7[b]	9.2[b]
Foreign exchange reserves ($ billion end year)	11.9	10.5	15.2	17.5	17.0	28.6

Sources: IMF, *International Financial Statistics*; World Bank, *World Debt Tables*; Economist Intelligence Unit.

a/ Customs data.
b/ Economist Intelligence Unit estimates, Global Forecasting Service.

Annex Table A-3. Financial statistics, 1985-1990
(Rmb billion)

	1985	1986	1987	1988	1989	1990
Budget surplus (deficit)	2.1	(7.0)	(8.0)	(8.0)	(9.2)	(15.0)
Deposits	427.3	538.2	657.2	742.6	901.4	1,164.5
of which:						
Urban savings	105.8	147.1	206.4	265.9	373.5	519.3
Rural savings	45.0	56.0	62.6	67.0	71.6	85.0
Loans	590.6	759.0	903.2	1,055.1	1,249.1	1,518.6
Interest rates						
Money supply	301.7	385.9	457.4	548.8	583.4	..
Foreign direct investment ($ billion)						
Contracted	5.9	2.8	3.7	5.3	5.6	6.6
Utilized	2.0	1.9	2.3	3.2	3.3	3.4

Sources: IMF, *International Financial Statistics*; State Statistical Bureau of China, *Statistical Yearbook of China*, various editions, press reports.

Annex Table A-4. Prices and wages, 1985-1990

	1985	1986	1987	1988	1989	1990
Retail prices (Percentage)	8.6	6.0	7.3	18.5	17.8	2.1
Urban retail prices	12.2	7.0	9.1	21.3	16.0	0.2
Rural retail prices	7.0	5.0	6.3	17.1	18.8	3.2
Farmers' income per head (Rmb)	397.6	423.8	462.6	544.9	602.0	630.0
Urban households' income per head (Rmb) a/	748.9	910.0	1012.2	1192.1	1387.8	1522.8
Urban average wage (Rmb) a/	1148	1329	1458	1747	1935	2140

Source: State Statistical Bureau of China, *China Statistical Yearbook* (various issues).

a/ Includes subsidies for meat price rises, bonuses, overtime, etc.

Annex Table A-5. Investment in fixed assets, 1985-1990
(Rmb billion)

	1985	1986	1987	1988	1989	1990
Total (by ownership)	254.3	302.8	364.1	449.7	413.77	444.93
State-owned capital	168.1	197.9	229.8	276.3	253.55	291.86
Capital construction	107.4	117.6	134.3	157.4	155.17	170.38
Upgrading	44.9	61.9	75.9	98.1	78.88	83.02
Other	15.7	18.3	19.6	20.8	19.50	19.91
Collective	32.7	39.2	54.7	71.2	57.00	52.95
Urban	12.8	14.6	18.1	25.5	18.56	16.34
Rural	20.0	24.5	36.6	46.7	38.44	36.61
Individual	53.5	64.9	79.6	102.2	103.23	100.12
Urban	5.7	7.5	10.1	15.7	14.02	12.47
Rural	47.8	57.5	69.5	86.5	89.20	87.65
(by source of funds)						
State budget	40.8	44.1	47.6	40.3	34.16	38.77
Domestic loans	51.0	63.8	83.6	91.5	71.64	87.09
Foreign investment	9.1	13.2	17.5	25.4	27.42	27.83
Own and other funds	153.4	180.9	215.4	287.5	280.56	291.25
(by purpose)						
Construction	165.5	199.3	237.8	290.9	281.26	296.28
Equipment	71.8	82.3	99.7	122.1	104.87	114.84
Other	17.0	20.4	26.6	31.7	27.65	33.81
Productive construction	154.4	183.4	229.2	282.9	257.20	276.83
Non-productive construction	99.9	118.0	134.9	161.8	156.58	168.10
of which:						
Residential	64.2	72.9	87.2	106.7	106.38	116.45

Sources: State Statistical Bureau of China, *China Statistical Yearbook* (various issues); official communiques and press reports.

Annex Table A-6. Investment highlights, 1989 and 1990
(Key State-owned units)

	1989	1990
I. Investment in fixed assets (Rmb 100 million)		
Total	**2535.48**	**2918.64**
Agriculture	62.16	80.29
Energy industry	705.64	846.74
Transportation, postal and telecommunication service	271.74	348.41
II. Capital construction projects (unit)		
Total		
Under construction	66382	68261
of which:		
New projects	29320	36308
Completed and put in operation	35370	36502
Agriculture		
Under construction	774	815
of which:		
New projects	459	585
Completed and put in operation	506	562
Energy industry		
Under construction	2489	2560
of which:		
New projects	765	868
Completed and put in operation	894	928
Transportation, postal and telecommunication service		
Under construction	3764	4073
of which:		
New projects	1626	1952
Completed and put in operation	1792	2068

Source: State Statistical Bureau of China.

Annex Table A-7. Output of major industrial products, 1949-1990, selected years

Industrial products	Unit	1949	1960	1970	1980	1985	1986	1987	1988	1989	1990
1.Food, beverages, tobacco											
Salt	10,000 tons	299.00	1287.00	1109.00	1728.00	1479.00	1761.00	1764.00	2264.00	2828.98	2022.64
Sugar	10,000 tons	20.00	44.00	135.00	257.00	451.00	525.00	506.00	461.00	500.90	581.95
Canned food	10,000 tons		11.80	19.30	57.20	142.50	164.10	161.50	220.90	232.52	157.14
Beer	10,000 tons		15.00	16.00	69.00	310.00	413.00	540.00	656.00	643.00	692.21
Cigarettes	10,000 cases	160.00	499.00	783.00	1520.00	2370.00	2596.00	2881.00	3096.00	3195.06	3297.53
2.Non durable consumer goods											
Textiles											
Chemical fibres	10,000 tons		1.06	10.09	45.03	94.78	101.73	117.50	130.12	148.09	165.42
Yarn	10,000 tons	32.70	109.30	205.20	292.60	353.50	397.80	436.80	465.70	476.72	462.58
Knitting wool	tons	0.18	0.95	2.17	5.73	12.59	14.91	20.47	22.50	25.00	23.80
Cloth	100 million metres	18.90	54.50	91.50	134.70	146.70	164.70	173.00	187.90	189.23	188.76
Woollen piece goods	10,000 metres	544.00	3646.00	5776.00	10095.00	21816.00	25187.00	26538.00	28609.00	27962.00	29504.04
Gunny bags	100 million	0.10	0.86	1.84	4.10	6.27	7.60	8.59	9.31	7.84	7.41
3.Durable consumer goods											
Sewing machines	10,000		88.00	235.20	767.80	991.20	989.40	970.00	983.23	956.30	760.96
Bicycles	10,000	1.40	176.50	368.80	1302.40	3227.70	3568.30	4116.70	4140.10	3676.84	3141.57
Wrist watches	10,000		50.50	347.60	2215.50	5431.10	7317.40	6142.40	6661.60	7275.60	8352.64
Fine aluminium products	10,000 tons		0.32	1.91	5.80	7.26	8.79	10.58	8.99	8.22	7.85
Bulbs	100 million	0.13	2.19	3.86	9.46	15.33	16.09	16.83	18.27	20.80	24.51
Household washing machines	10,000				24.50	887.20	893.40	990.20	1046.80	825.43	662.68
Electrical fans	10,000				723.70	3174.60	3528.70	3660.70	4495.50	4991.93	5799.33
Radio sets	10,000	0.40	158.70	323.10	3003.80	1600.30	1589.30	1763.80	1548.90	1834.72	2102.99
Television sets	10,000		0.79	1.05	249.20	1667.66	1459.40	1934.37	2505.07	2766.54	2684.70
Cameras	10,000		17.31	4.04	37.28	178.97	202.54	256.70	312.26	245.18	213.22
4.Intermediate goods											
Pig iron	10,000 tons	25.00	2716.00	1706.00	3802.00	4384.00	5064.00	5503.00	5704.00	5820.02	6238.32
Steel	10,000 tons	15.80	1866.00	1779.00	3712.00	4679.00	5220.00	5628.00	5943.00	6158.71	6634.83
Rolled steel	10,000 tons	13.00	1111.00	1188.00	2716.00	3693.00	4058.00	4386.00	4689.00	4859.08	5153.39
Cement	10,000 tons	66.00	1565.00	2575.00	7986.00	14595.00	16606.00	18625.00	21014.00	21029.47	20971.08
Plate glass	10,000 standard cases	108.00	670.00	1053.00	2771.00	5606.00	5202.00	5803.00	7293.00	8441.79	8066.59
Timber	10,000 cubic metres	567.00	4129.00	3782.00	5359.00	6323.00	6502.00	6408.00	6218.00	5801.77	5108.53
Soda ash	10,000 tons	8.80	81.50	107.70	161.30	201.10	215.40	236.30	260.90	304.22	379.51
Calcium carbide	10,000 tons	8.30	24.30	69.60	152.00	195.30	215.00	241.20	225.60	246.13	227.98
5.Capital goods											
Mining equipment	10,000 tons	0.07	25.19	9.63	16.25	31.43	29.06	29.72	38.36	32.56	31.38
Power generating equipment	1,000 kilowatt		338.80	291.80	419.30	563.00	722.40	941.10	1109.30	1174.03	1225.38
Metal cutting machine tools	10,000	0.16	15.35	13.89	13.36	16.72	16.37	17.22	19.17	17.87	13.45
Motor vehicles[a]	10,000		2.26	8.72	22.23	43.72	36.98	47.18	64.47	58.35	51.40
Tractors	10,000		1.16	3.19	9.77	4.50	2.86	3.71	4.72	3.98	3.94
Walking tractors	10,000		0.12	5.14	21.79	82.25	77.45	110.60	133.57	111.81	110.14
Locomotives			804.00	573.00	512.00	746.00	818.00	909.00	844.00	680.00	466.00
Railway passenger coaches		23.00	818.00	576.00	1002.00	1447.00	1522.00	1791.00	1980.00	2000.00	1866.00
Railway freight wagons	10,000	0.14	2.61	1.38	1.06	1.93	2.06	2.16	2.33	2.41	1.86

Source: State Statistical Bureau of China.

a/ Including passenger vehicle.

Annex Table A-8. Number of staff and workers in State-owned enterprises by subsector of manufacturing, 1960-1990, selected years
(10,000 persons)

Subsector of manufacturing	1960	1970	1980	1982	1985	1986	1987	1988	1989	1990
Food manufacture	}			}	206.1	213.2	219.7	242.6	}	
Beverage manufacture	}136.9	125.0	234.0	} 274.4	64.5	70.3	77.2	86.4	}356.8	353.4
Tobacco manufacture	}			}	21.3	22.5	23.8	26.9	}	
Fodder manufacture	}			}	4.0	7.1	7.3	9.4	9.8	10.9
Textile manufacture	}198.8	192.0	310.8	} 412.1	422.9	450.0	474.5	500.7	500.2	506.3
Chemical fibres	}			}	22.9	26.0	28.6	24.7	26.6	29.3
Clothing	..	..	9.2	15.7	16.7	16.5	18.0	19.1	19.2	21.2
Leather, furs and manufactured goods	..	..	16.1	20.9	21.2	22.3	22.9	24.0	23.0	23.6
Timber processing, bamboo, cane, palm fibre and straw products	..	..	25.3	29.2	25.8	26.9	25.9	26.2	26.1	26.2
Furniture manufacture	..	..	..	..	6.3	7.4	6.5	6.6	6.5	6.5
Paper making and manufactured goods	20.5	23.5 }	}	}	58.4	61.6	66.3	69.8	71.3	73.2
Printing	..	..	} 97.3	} 60.5	49.7	51.4	52.5	53.1	53.3	54.5
Cultural, educational and sports articles	..	..	..	..					10.1	10.9
Arts and crafts	..	..	..	..	5.9	6.7	7.1	8.0	8.2	8.9
Power generation, steam and hot water production and supply	25.0	33.3	81.3	95.0	116.4	123.6	133.1	142.6	148.4	156.9
Petroleum processing	16.4	24.3	16.8	22.2	29.7	33.2	35.4	41.3	42.2	44.3
Coking, gas, and coal related products	..	..	5.5	5.1	14.0	15.3	16.4	17.9	18.7	21.0
Chemical industry	}				259.1	268.0	284.6	300.2	304.5	311.0
Medical and pharmaceutical products	}			}	45.5	48.9	53.2	58.2	60.0	63.6
Rubber manufactured goods	}123.1	146.9	286.7	}	36.7	38.8	40.7	41.5	42.9	44.7
Plastics manufactured goods	}			} 294.2	18.2	19.1	20.7	20.9	21.3	23.2
Building materials and other non-metal mineral products	}	115.6	190.6	195.8	249.8	264.3	269.2	278.1	269.5	270.1
Smelting and pressing of ferrous metals	}167.4	}199.8	192.3	193.1	228.1	234.8	240.8	247.6	251.2	256.1
Smelting and pressing of non-ferrous metals	}	}	45.8	50.5	52.0	54.9	66.1	67.8	72.0	69.4
Metal products	}			}	50.6	51.0	53.7	53.1	52.5	54.3
Machine building	}			}	575.2	585.9	594.6	693.4	702.8	696.0
Transportation equipment	}			}	194.2	196.2	202.0	207.2	}	148.9
Electric equipment and machinery	}			}	97.6	102.6	108.4	114.6	}	96.2
Electronic and telecommunications equipment	}486.1	558.3	925.8	}1017.4	88.5	90.7	90.8	93.8	}363.6	99.7
Instruments, meters and other measuring equipment	}			}	44.7	45.7	46.1	47.3	}	41.9

Source: State Statistical Bureau of China.

Annex Table A-9. Selected performance indicators of State-owned independent accounting industrial enterprises, 1982-1990, selected years

Item	Unit	1982	1985	1988	1989	1990
I. Number of enterprises		63,063	69,834	72,494	73,501	74,775
II. Original value of fixed assets (year end)	Rmb 100 million	4,374.95	5,956.15	8,795.20	10,160.84	11,610.27
of which for industrial production	Rmb 100 million	3,590.24	4,870.11	7,126.72	8,259.20	9,469.04
III. Total funds	Rmb 100 million	4,145.92	5,604.06	8,603.36	10,318.00	12,356.26
Net value of fixed assets (year end)	Rmb 100 million	2,914.01	3,980.77	6,040.36	7,033.20	8,088.31
Quota circulating fund[a/]	Rmb 100 million	1,231.91	1,623.29	2,563.00	3,284.00	4,267.95
IV. Revenue from sale of products	Rmb 100 million	4,030.73	7,922.31	9,815.99	11,102.13	11,718.74
V. Net industrial output value (at current prices)	Rmb 100 million	4,146.66	2,057.73	3,062.98	3,460.23	3,540.02
VI. Profits and taxes	Rmb 100 million	972.19	1,334.14	1,774.87	1,773.14	1,503.14
Profits	Rmb 100 million	597.66	738.17	891.85	923.20	388.11
Taxes	Rmb 100 million	374.53	595.97	883.02	849.94	1,115.03
VII. Net output value per Rmb 100 Original value of fixed assets	Rmb	94.78	34.55	34.83	34.60	30.49
VIII. Profits and taxes per Rmb 100 Original value of fixed assets	Rmb	22.22	22.40	20.18	17.45	12.95
IX. Ratio of profits and taxes to total fund	Per cent	23.45	23.81	20.63	17.18	12.17
X. Ratio of profits and taxes to cost	Per cent	31.98	29.09	22.48	19.42	14.41
XI. Decrease rate of cost of comparable products[b/]	Per cent	-0.38	-7.70	15.59	-22.17	-7.03

Source: State Statistical Bureau of China.

a/ Calculated on basis of average holding of circulating fund of the year.
b/ Minus sign (-) indicates an increase in cost.

Annex Table A-10. China's major trade partners, 1990
(In thousand $)

Country	Total trade	PRC exports (f.o.b.) 1990	PRC exports (f.o.b.) 1990/89 (Percentage change)	PRC imports (c.i.f.) 1990	PRC imports (c.i.f.) 1990 (Percentage change)
Japan	1659901	901102	7.34	758799	-27.97
Hong Kong	4090772	2665006	21.60	1425766	13.69
United States	1176779	517946	17.45	658833	-16.21
Germany, Federal Republic of	497108	203439	26.46	293669	-13.09
Singapore	283217	197466	16.70	85751	-20.44
Canada	190872	43035	4.52	147837	37.16
Jordan	10936	7509	-76.47	3427	5.61
USSR	437911	223919	21.09	213992	-0.32
Australia	180865	45509	7.55	135356	-8.06
United Kingdom	202693	64304	1.24	138389	27.72
Brazil	62943	10668	26.31	52275	-44.39
Italy	190483	83503	16.84	106980	-41.71
Romania	44147	22677	-25.95	21470	-54.61
France	230843	64540	22.22	166303	17.09
People's Republic of Korea	48274	35816	-5.09	12458	-32.50
Netherlands	130737	90827	19.16	39910	-11.34
Thailand	119375	82345	17.65	37030	-51.04
Malaysia	117614	34079	-3.24	83535	20.65
Syria	13721	13292	186.59	429	5262.50
Belgium	65797	32676	30.90	33121	-11.70
Philippines	29513	21010	-18.29	8503	2.59
Pakistan	58486	49480	34.42	9006	-59.86
Indonesia	118226	37902	70.06	80324	37.93
Spain	50682	18681	18.16	32001	-5.94
Switzerland	57091	15969	-42.81	41122	-21.83
Poland	32194	7432	-80.58	24762	-32.10
Czechoslovakia	61729	34143	-9.10	27586	-47.11
Sweden	40456	14103	1.76	26353	-7.36
Cuba	57815	27194	28.12	30621	33.69
Egypt	9141	8546	22.56	595	-7.47

Source: Customs.

Annex Table A-11. Source of foreign investment[a/] in China, 1979-1990
(Contracted investment, $ million)

Source	1979-1983	1984	1985	1986	1987	1988	1989	1990
Total	**7452.6**	**2874.9**	**6333.2**	**3330.4**	**4319.1**	**6190.7**	**6294.1**	**6986.3**
Hong Kong, Macao	4319.2	2175.5	4134.3	1773.4	2364.7	4161.2	3734.0	4257.7
United States	859.8	165.2	1152.0	541.5	361.5	384.3	645.9	365.7
Japan	954.9	203.0	470.7	282.8	386.3	370.6	515.4	478.5
United Kingdom	321.4	12.6	44.3	51.7	28.6	56.4	33.3	121.2
Singapore	54.5	62.6	75.5	140.8	79.8	136.9	147.7	106.9
Federal Republic of Germany	36.6	105.2	20.3	55.6	139.6	69.2	164.7	47.3
France	212.2	0.3	49.9	11.9	73.8	32.9	17.6	12.7
Italy	102.9	10.1	24.5	91.6	19.4	22.9	63.3	10.5
Canada	65.6	0.0	8.7	91.0	34.3	39.5	49.1	15.6
Australia	87.0	4.2	14.1	31.6	47.4	17.4	83.7	17.8
Switzerland	22.2	0.1	0.7	25.0	57.1	38.7	11.8	27.8
Austria	..	..	..	90.1	0.9	3.4	9.4	6.5
Philippines	4.3	2.1	40.6	3.8	30.5	15.5	4.7	10.8
Thailand	2.5	23.3	14.6	13.2	4.5	41.7	56.8	42.5
Belgium	42.0	8.1	2.7	0.3	1.7	0.6	21.2	6.1
Denmark	2.5	0.5	3.6	42.6	0.6	0.5	...	...
Sweden	18.3	14.6	5.0	0.0	5.3	0.8	27.6	3.5

Source: MOFERT data and Business China; State Statistical Bureau of China.

a/ This includes all kinds of investment - equity and co-operative joint ventures, wholly foreign-owned enterprises, joint offshore oil exploration, leasing, compensation deals and processing and assembly.

Annex Table A-12. Investment in fixed assets in industry by province, 1989 and 1990 (Rmb 100 million)

Province/Region	Total		Industry	
	1989	1990	1989	1990
Coastal region				
Beijing	79.83	84.84	12.83	13.72
Tianjin	37.65	40.06	25.04	28.16
Hebei	58.68	62.08	36.72	39.63
Liaoning	106.46	105.97	65.09	74.81
Shanghai	112.45	108.54	70.13	70.64
Jiangsu	74.42	73.48	42.76	41.49
Zhejiang	41.51	43.95	20.88	21.57
Fujian	33.77	38.19	17.54	20.25
Shandong	85.62	94.35	59.94	67.75
Guangdong	143.46	172.71	63.84	96.76
Guangxi	20.77	21.29	9.93	9.81
Hainan	19.98	22.07	6.98	6.71
Central region				
Shanxi	50.31	58.52	35.10	43.31
Inner Mongolia	27.58	33.92	19.83	24.94
Jilin	29.07	29.12	16.75	17.89
Heilongjiang	63.70	67.07	38.97	45.49
Anhui	33.83	37.57	20.78	22.04
Jiangxi	23.10	26.01	14.75	16.77
Henan	54.14	52.28	41.11	35.91
Hubei	42.36	52.82	22.76	28.97
Hunan	34.66	39.29	19.88	23.51
Western region				
Sichuan	81.80	89.76	50.20	56.43
Guizhou	17.77	20.58	11.45	14.89
Yunnan	22.19	27.19	10.83	13.20
Tibet	4.26	6.52	0.89	1.65
Shaanxi	38.66	41.25	23.22	26.13
Gansu	25.77	29.04	16.95	19.23
Qinghai	12.55	13.44	9.30	8.71
Ningxia	7.92	10.42	5.18	6.89
Xinjing	32.00	48.00	20.11	34.25
Not elsewhere classified	135.42	153.49	12.78	21.10
Total	**1551.74**	**1703.81**	**822.48**	**952.60**

Source: State Statistical Bureau of China.

Annex Table A-13. Industrial enterprises and gross output value of industry by province, 1988-1990

Province/Region	Number of enterprises 1988	1989	1990 (10,000)	Gross output value (Rmb 100 million) (at current prices) 1988	1989	1990
National total	**8105965**	**7980690**	**795.78**	**18224.58**	**22017.06**	**23924.36**
Coastal region						
Beijing	32256	31901	2.78	268.96	708.97	734.68
Tianjin	27444	30219	2.80	523.86	635.22	679.93
Hebei	632060	585602	56.86	841.12	1026.08	1123.23
Liaoning	274251	243582	25.44	1304.84	1546.43	1606.93
Shanghai	34088	36002	3.49	1295.87	1515.35	1632.94
Jiangsu	652025	622573	63.02	2152.93	2507.42	2764.10
Zhejiang	330529	332534	33.14	1141.03	1333.88	1432.55
Fujian	178963	181760	17.85	388.85	488.96	531.48
Shandong	576975	570917	58.90	1455.24	1920.94	2201.48
Guangdong	453363	425787	43.77	1318.91	1647.24	1902.24
Guangxi	237147	240505	24.91	271.82	326.99	353.43
Hainan	18671	16544	1.67	32.15	38.97	44.32
Central region						
Shanxi	116502	139539	13.30	349.44	487.65	538.39
Inner Mongolia	116353	83711	8.66	231.63	241.72	259.48
Jilin	180337	202367	17.86	454.24	530.36	552.39
Heilongjiang	150754	144028	13.95	686.04	804.77	863.51
Anhui	571775	584618	55.33	518.19	627.72	670.33
Jiangxi	346279	348909	35.81	345.33	406.16	425.75
Henan	816947	860749	81.08	779.90	953.57	1036.73
Hubei	407029	388744	38.78	834.84	976.92	1008.19
Hunan	420176	405245	41.94	581.85	680.09	712.66
Western region						
Sichuan	774252	758223	78.14	964.10	1147.29	1222.95
Guizhou	219818	200494	19.39	166.81	201.45	218.16
Yunnan	170111	170854	17.17	244.63	304.91	345.25
Tibet	4721	5409	0.48	1.25	2.87	3.07
Shaanxi	198970	206461	22.30	331.74	406.71	442.58
Gansu	79475	77130	8.31	204.27	248.30	277.67
Qinghai	17349	19096	1.87	42.51	54.18	55.24
Ningxia	24604	26081	2.63	44.00	58.54	64.75
Xinjiang	42211	41106	4.20	149.09	187.41	219.93
Not elsewhere classified	18671		-	31.25	-	-

Source: State Statistical Bureau of China, *China Statistical Yearbook* (various issues).

Annex Table A-14. Physical output of selected food products, 1978-1990

Items	1978	1979	1980	1981	1982	1983	1984	1985	1986	1987	1988	1989	1990
Salt (million tons)	19.53	14.77	17.28	18.32	16.38	16.13	16.42	14.79	17.61	17.64	22.64	28.29	20.22
Sugar (million tons)	2.27	2.50	2.57	3.17	3.38	3.77	3.80	4.51	5.25	5.06	4.61	5.01	5.82
Cigarettes (million cases)	11.82	13.03	15.20	17.04	18.85	19.38	21.32	23.70	25.96	28.81	30.96	31.95	32.98
Canned food (1000 tons)	488	501	572	684	785	845	1090	1425	1641	1615	2209	23.25	15.71
Beer (1000 tons)	400	520	690	910	1170	1630	2240	3100	4130	5400	6560	6430	6921

Source: State Statistical Bureau of China, *China Statistical Yearbook* (various issues).

Annex Table A-15. Physical output indices of major textile and clothing items, 1980-1990
(1980 = 100)

Year	Yarn		Cloth		Knitting wool		Woollen good		Gunny bags		Silk		Chemical fibre	
	(10,000 tons)	Index	(100 metres)	Index	(10,000 tons)	Index	(10,000 metres)	Index	(100 million)	Index	(10,000 tons)	Index	(10,000 tons)	Index
1980	292	100	134	100	5.73	100	10095	100	4.1	100	3.5	100	45	100
1981	317	109	142	106	7.65	133	11308	112	4.2	102	3.7	106	52	116
1982	335	115	153	114	9.25	161	12669	125	5.0	122	3.7	106	51	113
1983	327	112	148	110	10.21	178	14291	142	5.5	134	3.6	103	54	120
1984	321	110	137	102	11.00	192	18049	179	5.4	132	3.7	106	73	162
1985	353	121	146	109	12.59	220	21816	216	6.2	151	4.2	120	94	209
1986	397	136	164	122	14.91	260	25187	249	7.6	185	4.7	134	101	224
1987	436	148	173	129	20.47	357	26535	263	8.5	207	5.1	146	117	260
1988	465	159	187	139	22.50	393	28609	283	9.3	226	5.1	146	130	289
1989	477	163	189	141	25.00	436	27962	277	7.8	190	5.2	149	148	329
1990	463	159	189	141	23.80	415	29504	292	7.4	180	5.7	163	165	367

Source: State Statistical Bureau of China, *China Statistical Yearbook* (various issues).

Annex Table A-16. Iron and steel products of State-owned smaller steelworks, 1987
(In thousand tons)

Province	Plant	Iron	Steel	Rolled steel	Products
Hebei	Handan Iron and Steel	840.3	847.6	577.5	LS,MS,SS,WR,MP
	Xingtai Iron and Steel	374.6	—	60.2	WR
	Chengde Iron and Steel	192.2	166.8	121.1	MS,SS,QT,ST
	Shijiazhuang Iron and Steel	232.1	133.1	123.5	MS,SS,QS
Shanxi	Linfen Iron and Steel	286.0	—	49.1	WT
	Changzhi Iron and Steel	281.1	283.0	125.3	SS,QS,ST,R
Inner Mongolia	Huhe Iron and Steel	—	—	106.5	SS,QS,WR
Liaoning	Xinfu Steel	220.2	232.5	251.0	LS,MS,SS,QS
	Beitai Iron and Steel	329.3	16.5	15.3	LS,QS
	Lingyuan Iron and Steel	144.2	178.3	164.1	LS,QS,SP,WT
	Shengyang Iron and Steel	—	80.2	41.3	MS,SS,QS
	Shengyang Steel Mill	—	—	150.8	MS,SS
	Shengyang Wire Rod Mill	—	—	121.3	WR
	Dalian Steel Mill	—	—	128.1	SS,WR
	Dalian Second Steel Mill	—	4.0	139.3	SS,QS,WR
	Jinxi Steel Tube Mill	—	—	99.1	WT
	Yingkou Steel Plate Mill	—	—	200.0	MP
Jinlin	Tonghua Iron and Steel	386.7	230.8	178.8	MS,SS,QS
Heilongjiang					
	Xilin Iron and Steel	110.7	160.4	71.7	LS
	Haerbin Steel	—	—	270.6	MS,SS,WR
Jiangsu	Nanjing Iron and Steel	485.8	410.8	310.2	MS,SS,QS,MP,SH,SI
	Wuxi Iron and Steel	—	154.3	245.0	LS,MS,SS,QS,WR,ST,R
	Xuzhou Iron and Steel	174.0	140.2	37.8	WS
	Suzhou Iron and Steel	202.1	—	153.4	LS,WR
	Nanjing Steel Mill	—	—	146.0	CB,WR,WT
Zhejiang	Hangzhou Iron and Steel	345.0	372.2	387.3	LR,LS,MS,SS,QS,SH,SI, SP,ST
Anhui	Hefei Iron and Steel	232.0	282.1	186,3	LS,MS,SS,QS,WR,SH,SI,SP, ST,WT
	Wuhu Iron and Steel	100.7	—	10.7	SS
Fujian	Sanming Iron and Steel	331.3	320.1	260.9	LR,MS,SS,QS,WR,ST
Jiangxi	Xinyu Iron and Steel	490.4	211.0	190.4	WR,MP
	Jiangxi Iron and Steel	—	565.4	424.5	MS,SS,QS,WR,SH,SI,SP,ST, R
	Nanchang Iron and Steel	91.1	87.4	195.0	MS,SS,QS,SP,WT,R
	Pingxiang Iron and Steel	250.1	180.9	—	—

(continued)

Annex Table A-16. (continued)

Province	Plant	Iron	Steel	Rolled steel	Products
Shangdong	Jinan Iron and Steel	760.0	603.0	436.2	MS,SS,QS,MP
	Laiwu Iron and Steel	549.3	308.0	143.6	LS,MS,SS,QS,SP,WT
	Qingdao Steel	—	441.9	370.7	MS,SS,WR,SH,SI,ST
	Jinan Iron Works	208.2	—	—	—
	Zhangdian Iron and Steel	182.0	—	59.9	SP,WT
Henan	Anyang Iron and Steel	731.2	688.0	547.9	MS.SS,QS,WR,MP,SH,SI,ST
	Luoyang Steel	—	95.6	122.7	WR
Hubei	Echeng Iron and Steel	388.1	256.6	294.4	MS,SS,QS,WR,SP,WT
	Xialu Iron and Steel	87.9	121.4	129.7	MS,SS,QS
	Yichang Bayi Iron and Steel	—	80.3	98.8	MS,SS
Hunan	Lianyuan Iron and Steel	475.8	449.7	328.4	LR,LS,MS,SS,QS,SP,WT,R
	Hengyang Steel Tube Mill	—	—	97.3	ST,WT
	Lengshuijiang Iron & Steel	202.5	—	—	—
Guangdong	Shaoguan Iron and Steel	313.2	408.4	399.3	LS,MS,SS,QS,WR,MP,R
	Guangzhou Iron and Steel	279.7	331.4	308.3	MS,SS,QS,WR,ST
	Guangzhou Steel Mill	—	—	135.7	WR
Guangxi	Liuzhou Iron and Steel	240.8	328.8	383.8	MS,SS,WR,MP,SH,SI,ST
Sichuan	Weiyuan Iron Works	144.9	151.0	97.7	MS,SS,WR
	Chengdu Iron and Steel	44.1	105.6	222.2	SS,WR
	Daxian Iron and Steel	—	115.6	88.6	SS
Yunnan	Kunming Iron and Steel	583.7	551.2	417.7	LS,MS,SS,QS,WR,MP,SH,SI, ST,R
Shaanxi	Lueyang Iron and Steel	158.2	74.6	48.6	SS
	Xian Iron and Steel	—	—	56.2	SS,QS,WR
Gansu	Lanzhou Steel	—	242.5	192.2	SS,QS,WR,SP,WT
Xinjiang	Bayi Iron and Steel	163.5	181.7	150.4	MS,SS,QS,WR,SP
Total		**11,623.0**	**10,639.5**	**10,539.8**	

Source: "Chinese Steel", *MBM* (December 1988), p. 63.

Note: LR - Light rails; LS - Large sections; MS - Medium sections; SS - Small sections; QS - all sections in "Quality" steel; CB - Cold "blendings"; WR - Wire rod; MP - Medium plate; SH - Sheet; SI - Silicon sheet; SP - Strip; ST -Seamless tube; WT - Welded tubes; R - remainder.

Annex Table A-17. Physical output of selected chemicals, 1981-1990

Year	Sulphuric acid ('000 tons)	Soda ash ('000 tons)	Caustic soda ('000 tons)	Effective components of chemical fertilizers ('000 tons)			Chemical pesticides ('000 tons)	Ethylene ('000 tons)	Calcium carbide ('000 tons)	Plastics ('000 tons)
				Total	Nitrogenous	Phosphate				
1981	7807	1652	1923	12390	9857	2508	484	504.8	1513	916
1982	8175	1735	2073	12781	10219	2537	457	564.9	1675	1003
1983	8696	1793	2123	13789	11094	2666	331	653.7	1808	1121
1984	8172	1880	2222	14602	12210	2360	299	648.0	1846	1180
1985	6764	2011	2353	13222	11438	1760	211	652.1	1953	1234
1986	7631	2146	2518	13957	11592	2340	203	695.2	2150	1321
1987	9833	2363	2739	16722	13423	3259	161	937.2	2412	1526
1988	11113	2609	3005	17402	13656	3692	179	1232.1	2256	1904
1989	11533	3042	3211	18025	14241	3728	208	1396.0	2461	2058
1990	11690	3795	3354	19969	14797	4255	229	1572.0	2280	2196

Source: State Statistical Bureau of China, *China Statistical Yearbook* (various issues).

Annex Table A-18. Physical production[a/] of selected petrochemicals, 1984-1988 (In thousand tons)

Item	1984	1985	1986	1987	1988
Ethylene	551	551	587	817	1111
Synthetic resins	485	498	555	728	1026
Synthetic rubbers	128	129	130	139	184
Monomers for synthetic fibres	359	487	505	525	589
Polymers for synthetic fibres	365	447	458	479	491
Synthetic fibres	241	308	314	329	347

Source: *World Chemical Industry Yearbook, China Chemical Industry* (1988 and 1990 Editions).

a/ Output of enterprises under the China Petro-chemical Corporation.

Annex Table A-19. Output of machine tools, 1984-1990

Year	Metal-cutting machine tools ('000)	of which: large scale ('000)	High precision ('000)	Numerically controlled machine tools ('000)	Output value of machine tools (billion yuan)
1984	133.5	3.907	1.621	1.613	4.3
1985	167.2	4.479	1.721	1.959	4.9
1986	163.7	4.682	1.618	2.393	5.4
1987	172.2	4.649	1.304	2.604	6.0
1988	191.7	4.860	1.475	2.681	7.0
1989	178.7	3.866	1.117	2.742	..
1990	134.5	2.295	1.121	2.634	..

Sources: State Statistical Bureau of China, *China Statistical Yearbook* (various issues).

Annex Table A-20. Production of automobiles, 1978-1989, selected years ('000)

Type/Year	1978	1983	1985	1987	1988[a]	1989[b]
Total domestic production	149.0	240.0	437.0	472.0	500.0[a]	650.0[b]
Truck	25.4	199.3	351.0	422.0	..	..
Jeep	13.9	18.2	20.7	46.0	..	..
Domestic car	2.6	6.0	5.7	4.0	..	..
Foreign car assembled in China				22.0	..	..
Imports			370.0	200.0	..	..

Source: China National Automotive Corporation (CNAIC).

a/ Preliminary estimates.
b/ Projection based on a plannned increase of 100-150,000 automobiles.

Annex Table A-21. Production of oil and gas, 1990[a]

Producing area	Oil 1,000 b/d	Percentage of annual plan	Gas Mcf/d	Percentage of annual plan
Daqing	1,124	58.4	218	58.9
Shengli	672	57.8	138	56.4
Huabei	109	59.3	24	68.5
Liaohe	273	58.1	171	59.8
Xinjiang	134	56.7	48	..
Dagang	77	54.8	34	56.2
Henan	50	56.7	4	..
Zhongyuan	127	53.0	136	59.7
Jilin	72	62.3	10	..
Changqing	30	61.0	3	..
Jianghan	17	57.7	7	..
Yumen	11	68.2	1	..
Jiangsu	17	64.5	4	..
Qinghai	15	41.8	4	..
Sichuan		75.8	636	61.0
Jidong	7	41.6	3	..
Offshore	20	56.7	..	..
Total	**2,755**		**1,441**	

Source: China National Petroleum Corporation.

Annex Table A-22. Degree of rural industrialization in Chinese Provinces, 1986

Most industrialized (Rural industry output more than 50 per cent of RSVO)			Intermediate stage of rural industrialization (Rural industry output 15-50 per cent of RSVO)			Least industrialized (Rural industry output less than 15 per cent of RSVO)		
	Rural industry Percentage RSVO	Rural industry Percentage rural employment		Rural industry Percentage RSVO	Rural industry Percentage rural employment		Rural industry Percentage RSVO	Rural industry Percentage rural employment
Beijing	50.0	28.2	Heilongjiang	16.3	5.0	Xizang	1.1	1.1
Zhejiang	5.9	19.6	Jilin	18.9	4.2	Xinjiang	5.9	2.0
Jiangsu	52.0	19.6	Hunan	19.0	4.8	Neimenggu	7.6	2.9
Tianjin	53.6	29.6	Jiangxi	19.9	6.3	Qinghai	7.9	3.0
Shanghai	65.6	45.8	Anhui	20.2	4.4	Guangxi	10.6	1.9
			Sichuan	20.4	4.3	Ningxia	11.7	1.9
			Hubei	22.6	9.1	Yunnan	11.8	2.3
			Shaanxi	24.5	6.9	Guizhou	13.1	5.9
			Guangdong	25.1	8.6	Gansu	13.1	3.2
			Henan	26.2	4.8			
			Fujian	27.6	5.7			
			Shandong	31.6	8.6			
			Hebei	38.4	9.1			
			Shanxi	41.0	11.0			
			Liaoning	41.1	13.1			

Source: Kenneth R. Walker, "Forty years on: Provincial contrasts in China's rural economic development", *The China Quarterly* (September 1989), p. 451.

a/ The rural social value of output (RSVO) is defined to include "agriculture" (which itself embraces crops, livestock, fisheries, forestry and subsidiary activities), industry, construction, transport, commerce, the food industry of *xiang* (townships) or levels below the *xiang*, for example, co-operative organizations and farm households, and also the value of agricultural output of State farms. It excludes the output of rural State-run industry, construction, etc. and that of State farms, and *xian* (counties) and *zhen* (towns).

ANNEX B

TECHNICAL NOTES ON CHINESE STATISTICAL TERMS AND CONCEPTS

Annex B. Technical notes on Chinese statistical terms and concepts

These technical notes on Chinese statistical terms and concepts are provided to assist the reader who is unfamiliar with Chinese statistics. The statistical system of China was originally designed to serve the needs of government economic planners and data are frequently divided into categories that reflect administrative needs. More recently, the collection and reporting of statistical data in China has undergone adjustments in response to the reform of China's economy. The concepts and terms explained below are provided to clarify basic terms.

Basic terms

Ownership

(1) State-owned units (including government agencies, also referred to as units owned by the whole people)
(2) Collective-owned units
(3) Individual-owned units (refer to units owned by one or more Chinese individuals)
(4) Joint ownership units:
  (a) State and collectives
  (b) State and individuals
  (c) Collective and individuals
  (d) Chinese and foreign investors
(5) Foreign-owned units:
  (a) Overseas Chinese or Chinese business persons from Hong Kong and Macao
  (b) Foreign investors

China's statistical data are often classified by type of ownership. For statistical reporting, categories 4 and 5 are generally combined into one category called Other Ownership Units.

Material production vs. non-material production

(1) Material production includes five sectors:
  (a) Agriculture
  (b) Industry
  (c) Transportation
  (d) Construction
  (e) Commerce
(2) Non-material production sectors include all kinds of services other than commerce. Note that commerce is statistically classified as a material production sector.

Productive vs. non-productive investment

An investment project is classified as productive or non-productive according to whether it is directly related to material production.

(1) Productive projects refer to those which increase material production or that directly serve material production.
(2) Non-productive projects refer to residential buildings, public services and facilities for scientific research, culture, education, and health care.

Agriculture

(1) Farming (cultivation of crops)
(2) Forestry

(3) Animal husbandry
(4) Sideline production: collection of wild plants, hunting of wild animals and fowl, handicrafts and small-scale industries concurrently run by farmers
(5) Fishery

Industry

(1) Extraction of natural resources, including mining and lumbering
(2) Processing of agricultural products
(3) Manufacture of industrial products
(4) Repair of capital goods
(5) Electricity generation and supply, water purification and gas production

Society

Indicators may be referred to as "society" or "social", e.g., total investment of society, social labour force. The addition of the words "social" or of "society" means that data include all forms of ownership. For example, "Social labour force" includes all individuals employed and self-employed by all kinds of ownership units in urban and rural areas.

Administrative ranking of city

A city typically consists of the city proper (*shiqu* or *shixiaqu*) which is officially designated as urban and some rural countries under its administration (*jiaoxian* or *shixiaxian*) which may be predominately agricultural.
China's administrative structure has four basic levels: central government (*zhongyang*), province (*sheng*), prefecture (*diqu*), and country (*xian*). Cities are also ranked according to this hierarchy which determines political and economic autonomy. For example, Beijing is equivalent to a province for administrative purposes and reports directly to the central government. Shangahi and Tianjin are also "provincial" level cities. Below the provincial level there are 170 prefectural level cities reporting directly to the provincial governments. These include provincial capitals and other major cities. Finally, there are county level cities which report to prefectures. Entire rural counties can become country level cities when the "non-agricultural" (*feinongye*) population reaches eighty per cent of the total population. Some special provisions also apply for border counties and those adjacent to special economic zones.

Planning period

The years since 1949 are grouped into the following planning periods: Rehabilitation period 1950-1952; the first Five-Year Plan period 1953-1957; the second Five-year Plan period 1958-1962; the Adjustment period 1963-1965; the third Five-Year Plan period 1966-1970; the fourth Five-Year Plan period 1971-1975; the fifth Five-Year Plan period 1976-1980; the sixth Five-Year Plan period 1981-1985; the seventh Five-Year Plan period started with 1986.

Prices

There are mainly three kinds of prices:

(1) List prices (Plan prices), which are determined by government agencies
(2) Negotiated prices, which State and collective owned units can adjust within limits, usually upper limits, set by government agencies
(3) Free market prices, which are determined by market supply and demand (government agencies may also set upper limits for free market prices)

Product and income statistics

Gross output value

The sum of output value (list price x quantity produced) of each enterprise in a given sector. Note that in summing the output value of enterprises to get gross output value of a sector, the value of intermediate goods is double counted. The sum of gross output value of agriculture, industry, transportation, construction, and commerce, the five material production sectors of the economy, is defined as total output value of society.

Net output value

The sum of value added of each enterprise in a given sector. Note that net output value does not double count value.

National income

The sum of net output value of agriculture, industry, transportation, construction, and commerce, the five material production sectors of the economy. Note the coverage of national income in Chinese statistics excludes the value added in "non-material production sectors". China's national income is approximately equivalent to the United Nations" net material product (NMP).

National income produced vs. national income used

The above defined national income is also known as national income produced. The national income used is defined as consumption plus accumulation (investment). The relationship between the two measures is national income used = national income produced - exports + imports.

Production material consumption

The difference between total output value of society and national income is production material consumption. It includes the value of all producer goods consumed, i.e.,, depreciation of fixed assets and materials, fuels, etc., in material production.

Gross material product

The sum, of value added for all sectors, "material production" and "non-material production" of the economy and depreciation of fixed assets.

Breakdowns of gross output value of industry

(1) Ownership

Three categories of ownership are commonly used:
- (a) State-owned industry
- (b) Collective-owned industry
- (c) Other ownership industry

Additional categories, such as individual industry in urban and rural areas, township and/or village run industry, are also used. However, adding these categories can create overlapping. For example, township and village run industry involves several kinds of ownership, collective, joint and individual ownership.

(2) Light and heavy industry
- (a) Light industry produces consumer goods and hand tools
- (b) Heavy industry produces capital goods

Gross output value of industry (note changes in coverage)

From 1958 to 1984, gross output value of industry included enterprises run by townships, formerly people's communes. Gross output value of agriculture included enterprises run by villages, formerly production brigades and teams. Since 1984, village run industries were transferred from agriculture to industry.
Township and village run industries are often combined into one category. However, this category includes several kinds of ownership, collective, joint, and individual ownership. In rural areas, for cotton ginning, rice husking, wheat milling, animal slaughtering, and sewing, only processing charges instead of the whole value of products are counted in gross output value of industry.

Gross output value of agriculture (note changes in coverage)

Before 1958, China's gross agricultural output value included barnyard manure, and handicraft products for self-consumption (clothes, shoes, stockings, and initial grain processing made or undertaken by farmers). Since 1958, forestry included the cutting and felling of bamboo and tress by units at the village level and below; the value barnyard manure was excluded from animal husbandry; sideline occupations included self-consumed handicrafts. But, the value of output of industries run by units at village level and below is included in sideline occupations and the output value of fish caught using motor fishing boats was added to fishery. Since 1980, the value of handicraft products made for sale by individuals has been added to sideline occupations. From 1984, industries run by units at the village level and below have been included in the sector of industry.

Gross output value of rural industry

Gross output value of rural industry refers to the total output of products, expressed in value terms, produced by enterprises of township industry (previously commune industry), village industry (previously brigade industry) and industry below village level (previously production team industry, joint sponsored industry by farmers, and individual industry). It includes both the value of finished products and the value of industrial operation services provided to other enterprises. Also included is the value of semi-finished products.

Employment statistics

Agricultural and non-agricultural residents

The status of agricultural and non-agricultural residents is determined at birth according to household registration regulations. Generally, agricultural and non-agricultural residents live in rural or urban areas respectively, but this distinction may not reflect the nature of an individual's residential location or occupation. The status is largely related to access to State rations. Non-agricultural residents have certain privileges such as access to food coupons which allow them to purchase food at government subsidized prices. This privilege is retained when a non-agricultural residents moves to a rural area. Similarly, many agricultural residents may reside in a city but their status will remain agricultural. In recent years as cities have annexed increasing number of rural counties, the percentage of agricultural residents in the urban population has also increased.

Classifications

The employment statistics can be classified by type of ownership and by urban and rural locations. Eight categories are distinguished:

(1) Employees of State-owned units in urban areas (including government agencies)
(2) Employees of State-owned units in rural areas (including government agencies)
(3) Employees of collective-owned units in urban areas
(4) Employees of collective-owned units in rural areas

(5) Individual labourers in urban areas
(6) Individual labourers in rural areas
(7) Employees of other ownership units in urban areas
(8) Employees of other ownership units in rural areas

Individual labourers include individuals who own businesses, usually small-scale enterprises, and their employees.

Social labour force

Social labour force refers to all individuals employed or self-employed in urban and rural areas, including all eight categories in the above classification.

Staff and workers

Staff and workers is a literal translation of the Chinese term *zhigong*, which includes categories 1, 2, 3 and 7, and in addition to the above classification, figures for State-collective joint ownership include joint enterprises with township and village collectives, but figure for China-foreign joint ownership and foreign ownership exclude foreign employees. Foreigners include residents of Hong Kong and Macao.

Collective and individual labourers in rural areas

Collective and individual labourers in rural areas include categories 4, 6 and 8 in the above classifications.

The coverage can be otherwise specified as:

(1) Persons working under the responsibility system in collective units, the majority of whom are farmers
(2) Individuals employed by township and village run industries
(3) Individuals employed by rural units jointly owned by State and collectives, State and individuals, collectives and individuals, Chinese and foreign investors, or units wholly owned by overseas Chinese, Chinese business persons from Hong Kong or Macao, or foreign investors, in rural areas
(4) Individual labourers, individuals who own businesses and employees of those businesses

Number of persons newly employed in urban areas

Number of persons newly employed in urban areas refers to all persons who have registered with the labour bureau and have found jobs, and secondary school graduates who have registered with their street committees and found jobs, during a given period.

Investment

Investment in fixed assets

Total investment in fixed assets (of society) includes investment made by units of all kinds of ownership. For State-owned units, total investment is divided into three categories. Two major components are capital construction and technical updating and transformation, a third category, other investment in fixed assets, is used to refer to investment not covered by the first two categories. In principle, capital construction covers new construction and expansion projects, whereas technical updating and transformation covers renewing, replacing and rebuilding of existing assets.

Capital construction investment

Capital construction investment in new projects, including construction of a completely new facility, or an addition to an existing facility. It includes construction of plants, mines, railways, bridges, harbours, water conservation facilities, stores, residential facilities, schools, hospitals, and purchase of machinery and equipment, vehicles, ships and planes.

Technical updating and transformation

Technical updating and transformation investment refers to investment in projects to renew, modernize or replace existing assets.

Other investment in fixed assets

Other investment in fixed assets includes oil-field development projects using special oil-field development funds, development and expansion projects of mining and forestry department using maintenance funds, road and bridge construction projects using road tolls, warehouse construction of commerce department using simple construction funds and miscellaneous construction and purchase of fixed assets valued at 20,000-50,000 yuan.

Price indexes

Constant price

The average price of a given product specified by the State Statistical Bureau, to be used to calculate certain indicators, e.g., gross output value of industry. The indicators for different years at the same constant prices can be compared directly. Since 1949, the State Statistical Bureau has issued nationally unified constant prices four times: the 1952 constant prices for 1949-1957 period; the 1957 constant prices for 1958-1970; the 1970 constant prices for 1971-1980; the 1980 constant prices for the period since 1981. For comparison of an indicator at different constant prices, e.g., gross output value of 1954 and 1984, conversions are needed.

Comparable prices

Comparable prices are used to calculate index of certain indicators, e.g., national income.
Comparable prices include:
(1) Constant prices
(2) Current price divided by relevant price index.

Retail price indexes

There are two kinds of retail price indexes:
(1) Retail price index of society
(2) Retail price index of State-owned commerce

Both of them cover all consumer goods and the agricultural producer goods sold to farmers, exclude services and the agricultural producer goods sold to State-owned units.

Different kinds of prices are used in calculating retail price indexes. Retail price index of society, often referred to as overall retail price index, is calculated using all three kinds of prices, i.e., list prices, negotiated prices and free market prices. Since different kinds of prices are used, the index is a weighted average of each price weighted by the proportion of total expenditure on the good in question. Retail price index of State commerce is calculated using list prices only.
Retail price indexes are based on periodical sample survey. The sample set in 1984 includes 300 goods in more than 180 cities and 330 goods in more than 190 county towns.

Cost of living indexes

There are three kinds of cost of living indexes:
(1) Cost of living index of staff and workers
(2) Cost of living index of rural labourers
(3) Overall cost of living index (which cover both urban and rural areas)

Cost of living indexes cover consumer goods and services, which are considered necessities for daily life as determined by the State Statistical Bureau. Currently, 325 goods and services are selected for calculating cost of living indexes. Unless otherwise specified, cost of living indexes are based on list prices, negotiated prices, and free market prices.

Free market price index

Free market price index of urban areas covers consumer goods only, and free market price index of rural areas covers consumer goods and agricultural producer goods.

Price parity of industrial and agricultural products

The ratio of the prices of industrial products, Pi, and that of agricultural products, Pa. The parity can be expressed as either Pi/Pa or Pa/Pi. The parity can also be expressed in terms of the amount of industrial products for which a given amount of agricultural products can be exchanged.

Grassroots institutions

Rural (new) economic association

This is a translation of the Chinese term *Nongchun (Xin) Jingji Lianheti*. Since 1979 farmers have joined voluntarily economic organization that manage producing, processing and selling farm crops, and other business. The members of these associations jointly manage the associations and share the costs and benefits.

Township enterprises (Xiangshen qiye)

Township enterprises include:

(1) Enterprises previously run by communes and production brigade
(2) Co-operative enterprises run by rural labourers
(3) Individual-run enterprises and enterprises of other ownership at or below the township level

It was decided in March 1984 that all the above-mentioned enterprises should be called "township enterprises". Data in this Review include these kinds of enterprises at two levels: township level and village level.

Social institutions

Social institutions refer to all kinds of organization, e.g., government agencies, enterprises, army units and schools.

Large, medium and small enterprises

There are two criteria for distinguishing between large, medium and small enterprises:

(1) The annual production capacity of the enterprises. For enterprises that produce homogenous products, such as electricity, coal, petroleum, iron and steel, non-ferrous

metals, sulfuric acid, caustic soda, soda ash, synthetic ammonia, power-generating equipment, motor vehicles, tractors, timber, cement, plate glass, textiles, paper, sugar, wrist-watches, sewing machines, bicycles, etc., the production capacity is used as the criterion to decide whether the enterprises belong to the large, medium or small category. (For enterprises producing several products, the production capacity of the major product is used.)

(2) For enterprises which produce a variety of different products and are therefore not eligible for the above production capacity criterion, the size of the original value of their fixed assets is used to decide the size of the enterprises.

Industrial enterprises with independent accounting systems

An industrial enterprise can be classified as "with independent accounting systems", if its possesses independent administrative management and independent financial accounting.
The following criteria are necessary for an industrial enterprise to be considered as an independent accounting systems: possession of an independent administrative organization; ability to account independently for its profits and losses and prepare independent financial balance sheets; power to sign contracts with other units and control of an independent bank account. Any industrial enterprises with independent accounting, either single or combined, is to be treated as a basic reporting unit in preparing statistics.

Trade

Commodity purchasing power

Commodity purchasing power refers to the ability of individuals and social institutions to buy consumer goods and agricultural producer goods at retail markets. It includes three parts:

(1) Households' consumer goods purchasing power: total household income minus non-commodity expenditures (e.g., housing rents, medical cares)
(2) Social institutions' consumer goods purchasing power: budget of social institutions that can be used to buy consumer goods
(3) Farmers' agricultural producer goods purchasing power" farmers' income minus living expenditures and non-commodity production expenditure

Total value of retail sales of commodities

Total value of retail sales of commodities refers to the total value of consumer goods sold directly to individuals and social institutions, and the agricultural producer goods sold to farmers, excluding all other produced goods.

Trade grain

The quantity of purchases and retail sales of grain is calculated in terms of trade grain. Trade grain is unhusked rice and millet which have been converted into husked rice and millet according to standard conversion factors, while other kinds of grains remain in their unprocessed forms.

Exports and imports

The total value of imports and exports at customs refers to the value of commodities imported into and exported from China. This includes the actual imports and exports by foreign trade, imported and exported commodities under the processing and assembling trades and the imported and exported commodities and office utilities for Sino-foreign joint venture enterprises, contractual joint ventures and wholly foreign-owned enterprises. Also included are materials, supplies and gifts as aid given gratis between governments and by the United Nations and other international organizations and governments and contributions denoted by overseas Chines, compatriots in Hong

Kong and Macao and Chinese with foreign citizenship. Imports are calculated on a c.i.f. basis, while exports are on a f.o.b. basis.

Utilization of foreign capital

Utilization of foreign capital refers to loans, direct foreign investment, commodity credits and other funds used by domestic institutions that are supplied from abroad and from Hong Kong and Macao.

Contracted projects with foreign countries

Contracted projects with foreign countries refers to foreign construction projects undertaken by Chinese contractors, turnkey projects supplied through Chinese foreign-aid programmes, construction projects of Chinese organizations stationed abroad, technical assistance in the form of chargeable service, and complete plants and construction materials and supplied provided by Chinese contractors.

Service co-operation with foreign countries

Service co-operation with foreign countries refers to labour services provided to foreign employers by Chinese engineers, workers and management personnel, sent by Chinese international contracting corporations to work in foreign countries. The business income of labour service co-operation is the income of wages and salaries and other charges collected from the employers during the reference period.

Others

Total energy production

Total energy production refers to the total production of primary energy by all the energy industrial enterprises in a region in a given period of time. It includes raw coal, crude oil, natural gas, hydroelectricity and electricity generated by other means such as wind power and geothermal power, excluded bio-energy, solar energy and secondary energy converted from the primary energy.

Total energy consumption

Total energy consumption refers to the total energy consumption in production and living in a region in a given period of time. It includes raw coal and crude oil and their products, natural gas, and electricity, but exclude bio-energy and solar energy. Total energy consumption is divided into three categories:

(1) Final energy consumption

(2) Loss during the process of energy conversion: total input of energy of various forms for conversion, minus the total output of energy of various forms, minus petroleum products, minus other coke and chemical products in the country in a given period of time. It is an indicator of the loss that occurs during the process of energy conversion.

(3) Loss: total loss of energy due to managerial mistakes or any objective reasons during the course of energy production, transport and storage in a given period of time. The loss of various kinds of gas due to gas discharges and stocktaking is excluded.

Government revenue

(1) Revenues from enterprises, includes the profits if State-owned enterprises and the income of institutions delivered to government.

(2) Taxes include product tax, value added tax, resource tax, enterprise income tax, salt tax, custom duties, agricultural tax, husbandry tax, tax on slaughtering animals, tax on livestock trading, free market transaction tax, and fine and overdue taxes.

Income available

The income of households that can be used for daily expenses, i.e., total income minus expenditures to support senior family members not living together and taxes paid.

Source: State Statistical Bureau of the People's Republic of China, *China Statistical Yearbook 1989*, (Beijing 1990), pp. XXX1-XXXVIII.

ANNEX C

TWENTY KEY PROJECTS UNDER CONSTRUCTION

Annex C. Twenty key projects under construction

There are currently several hundred key projects under construction. The following 20 projects are representative of agriculture, energy, transport and communications, raw material and other basic industries.

1. Large grain production centres

China has started the construction of two large grain production centres. One, the Huanghe-Huaihe-Haihe Plain comprehensive development project, was begun in 1988. The project involves 2.4 billion yuan of investment contributed jointly by the State and relevant localities and is expected to treat 1.6 million hectares of low- and medium-yielding farmland, reclaim 90,000 hectares of wasteland and increase grain output by 5.5 billion kg. The other is the Huanghe (Yellow) River delta grain and cotton production centre. The project, started in 1989, involves 60 million yuan of investment by the State and localities. Upon completion, the project will divert some 130 million cubic metres of water from the Huanghe River, store 30 million cubic metres of water, increase grain output by 300 million kg and cotton output by 10,000 tons, and create conditions for the future large-scale, comprehensive agricultural development of the whole area.

2. The second phase of the "Sanbei" Shelter Belt Project

The first phase of the project was completed in 1985. The second phase, which began in 1986, is expected to be completed in 1995. With a total investment of 2.4 billion yuan, it will encompass 6.4 million hectares of newly established forests, 170,000 hectares of forests sown by air and 1.54 million hectares of forests established by halting the increasing drift of sand and closing the hill-sides off from livestock grazing and wood harvesting. In the past four years, 700 million yuan of investment has been used.

3. Flood control works on the lower reaches of Huanghe

The flood control works now under construction on the lower reaches of Huanghe are centred on two projects. One, the Guxian Reservoir on Luohe, a tributary of Huanghe, has a storage capacity of 1.15 billion cubic metres and will be able to cut the flow during the peak flood time at Huayuankou near Zhengzhou by 5,500 cubic metres per second to prevent a catastrophic flood. Involving a total investment of 600 million yuan, the project is expected to be completed by 1992. The second project involves raising and reinforcing the dyke on the lower reaches of Huanghe in order to ensure the safety of the surrounding areas and maintain the river's present tranquil state.

4. Large coal production bases

a) The Datong mining area in Shanxi Province

The coal pits under construction will have an annual capacity of 8.45 million tons. When completed, the mining area will play an important role in maintaining the sustained development of China's coal industry.

b) The Gujiao mining area in Shanxi Province

The project, with a designed annual production capacity of 16.5 million tons of raw coal and involving a total investment of 3.5 billion yuan, covers five pairs of shafts in Xiqu, Zhenchengdi,

Malan, Donggu and Tunlan. Except for the Tunlan mine, work on the other four pairs of shafts will be completely funded by energy loans from the Government of Japan. Upon completion, the developed mining area will greatly ease the shortage of coking coal for the iron and steel and chemical industries.

c) The Jungar coalfield in Inner Mongolia

Located in the eastern part of Jungar Banner in Ih Ju League, Inner Mongolia, the coalfield has an exploration area of 1,365 square kilometres and total verified geological reserves of 26.8 billion tons. The reserves' thick layered, shallow buried coal makes it well suited for open-cut mining. Moreover, the coal is of excellent quality, ideal for power generating and, because of its low sulphur and medium ash content, it fires with a high flame. The first phase of the projects involves the comprehensive development of coal, electricity and transport and includes the construction of a coal mine with an annual capacity of 15 million tons, a 215 km long railway from Fengzhen to Jungar with an annual transport volume of 15 million tons, and a coal pit mouth power station with an installed capacity of 200,000 kW.

5. The comprehensive development of Daqing oilfield

To ensure a steady, high output from the oilfield, the government has invested heavily during the current Seventh Five-Year Plan period (1986-1990) in prospecting and development work. By the end of 1990, the number of producing oil wells was expected to be 120 per cent more than that of the late 1980s. The additional crude oil annual exploitation capacity will amount to 19 million tons, helping to keep the oilfield's annual output at more than 55 million tons.

6. Large power stations

a) The Yantan hydroelectric power station

Construction of the project, with a combined installed capacity of 1.21 million kW and an annual output of 5.37 billion kWh of electricity, began in 1984. The first generating set is planned to begin operation in 1993 and the remaining three by 1995. When fully operational, the project will become the first 1 million kW power station in the Guangxi power grid; it will be incorporated into the south China electricity network and serve both the Guangxi Zhuang Autonomous Region and Guangdong Province.

b) The Geheyan hydroelectric power station in Hubei

This project is the first terraced hydropower station on the Qingjiang River, the second largest tributary on the lower middle reached of Changjiang (Yangtze) River. The first generating set is expected to start operation in 1993; the project is scheduled to be completed by 1995. Apart from easing the strain on the power supply in Hubei Province and the rest of central China, the project will serve as an important power load and frequency regulation station, helping to improve the quality of power supplied by the central China network and increasing the safety and economic return of the power grid.

c) The second stage of the Harbin No. 3 power plant

With an installed capacity of 1.2 million kW, it is equipped with two 600,000 kW generators manufactured by local State-owned enterprises using advanced imported technology. Construction of the project began in 1989 and is expected to begin generating electricity in 1993. By then, it will be able to supply sufficient electricity to the Daqing oilfield and Heilongjiang Province and ease the power shortage in the network of northeast China.

d) The Shentou No. 2 power plant in Shanxi

The project, a power station at a large coal mine, is 25 km away from the Pingshuo open-cut mine. The project began in 1986 and is planned for completion and operation by 1992. With an annual generating capacity of 6 billion kWh, it will promote the development of the Shanxi energy base and ease the strain on the power supply of the north China network.

e) The Daba power plant in Ningxia

The project, started in 1987, has a total investment of 700 million yuan contributed jointly by central and local governments and an installed capacity of 600,000 kW. It is scheduled to be completed by 1991. It will help raise the installed capacity of the Ningxia power grid from the present 930,000 kW to 1.53 million kW and be an additional source of power for the northwest China power network.

f) The Changshu power plant in Jiangsu

Located in the economically developed Jiangsu Province, the project is funded jointly by the State and the local government with a total investment of 1.3 billion yuan. Equipped with four 300,000 kW domestically produced generators, the project began in 1989. It is expected to begin generating electricity in 1993 and be completed by 1995. It will increase the power supply in southern Jiangsu Province and ease the power shortage in east China, particularly in the Changjiang River delta.

7. The Qinshan nuclear power station

Located at the northern foot of the Qinshan Mountain in Haiyan County, Zhejiang Province, this is the first 300,000 kW pressure water reactor nuclear power station designed and constructed by China. The project represents an important breakthrough in the country's peaceful use of nuclear energy. The design of the station is based on both China's national conditions and overseas standards. The project began in March 1985 with a total investment of 1.2 billion yuan and is expected to be completed by the end of this year. During project construction, a technical contingent has been brought to the fore which is capable to design, build and manufacture equipment for China's nuclear power industry. The project has laid a foundation for similar projects.

8. Datong-Qinhuangdao railway

The 652 kilometre long railway, from Shanxi Province's Datong to Hebei Province's Qinhuangdao Harbour, is China's first heavy-load special coal transport line. The project involves 6 billion yuan in investment. Construction was begun in the latter stage of the Sixth Five-Year Plan (1981-1985). The first stage, a 410 km double-tracked electrified line from Datong to Beijing, was completed and started operation at the end of 1988; it transported 20.15 million tons of coal in 1989. The second stage, with a length of 242 km from Dashizhuang to Qinhuangdao, is expected to be completed by 1992.

9. Third phase of the Qinhuangdao harbour coal wharf

The project is China's first modern coal wharf to unload coal directly from freight trains to cargo ships. It is China's largest coal export wharf. Two 35,000 ton berths and one 50,000 ton berth. with a combined loading capacity of 30 million tons, were completed and put into service in 1989. Construction of the whole project, with a total investment of 650 million yuan, will wind up this year. During construction, the foreign contractor claimed bankruptcy and unilaterally terminated the contract with the Chinese partner, adding great difficulties to the construction of the project. Displaying the spirit of self-reliance and hard struggle, the Chinese contractors in charge of construction, design and equipment manufacturing did their best to overcome the difficulties, fulfilling the task on schedule. The quality of engineering work was more than satisfactory with a 14 per cent savings in costs and additional savings in steel material.

10. The Beijing-Tianjin-Tanggu expressway

The project, a trunk highway linking Beijing with Tianjin and Tanggu new harbour, extends 142 km. The expressway will be monitored by a close circuit TV system and supported by automatic telephone and other service facilities. It is designed for a maximum passing speed of 120 km per hour. International bidding was invited for the construction of the project, and it is the first project in the history of China's highway construction that adopted the World Bank's construction management system. The World Bank held responsibility for engineering supervision of the project. The project involves a total investment of 1 billion yuan, including $150 million loan from the World Bank. The project, started in 1987, is expected to be completed and open to traffic in 1992.

11. Expansion of the Beijing telephone networks

The project is the largest of its kind in China in terms of the capacity of inner-city telephone exchange imported. There are also plans to introduce from abroad a 169,000 channel programme-controlled telephone exchange and a 4,700 channel long-distance telephone exchange. the completion of the project in 1991 will bring the city's telecommunications technology up to the advanced level of the 1980s and increase the telephone exchange capacity of Beijing to 500,000 channels, including 280,000 channels of programme-controlled exchanges, or more than 70 per cent of the total, and 6,700 channels of long-distance automatic telephone exchanges.

12. Second phase of Shanghai Baoshan Iron and Steel complex

the first phase of the project, completed and put into operation in September 1985, has an annual production capacity of 3.12 million tons of crude steel and involves a total investment of 12.877 billion yuan, including $2.78 billion in foreign exchange. The equipment is primarily from abroad (only 12 per cent is China made). The second phase of the project is still under construction but the plant's cold continual rolling and continuous casting workshops have already been completed. The installation of the furnace, sintering and coking equipment is well under way. The total investment involved is 17.24 billion yuan, including $2 billion in foreign exchange to import equipment. Upon completion, the second-phase project will have an annual capacity of 6.5 million tons of pig iron, 6.71 million tons of steel, 4 million tons of hot rolled steel plate, 2.1 million tons of cold rolled steel plate and 500,000 tons of seamless steel pipes. Some of these products are currently in short supply and have to be imported. The project will therefore help save foreign exchange otherwise used to import high-grade steel products and update China's production of steel in terms of output, variety and quality.

13. Expansion project of the Jinchuan Non-Ferrous Co.

The first-phase expansion project, completed in 1985, has an annual production capacity of 20,000 tons of nickel. Construction of the second phase, involving a total investment of 1.35 billion yuan, began in 1986. There are plans to dress 37,000 tons of nickel and copper, smelt 30,000 tons of these metals, and produce 220,000 tons of sulfuric acid and a small amount of platinum, palladium and gold. The completion of this project will greatly raise China's self-sufficiency rate for nickel and will bring its mining, dressing and smelting technology up to the advanced international level of the 1980s.

14. First phase of the Qinghai Potash Fertilizer Plant

The Qinhai Potash Fertilizer Plant in Golmud is the only large potash fertilizer factory in the country. Construction of this project, using an investment of 466 million yuan, began in 1986. Major equipment for the plant is being installed and some are in trial operation. The first phase of the project is expected to be finished in 1991. By then the plant will supply the State 200,000 tons of potassium chloride each year, saving the State $20 million in foreign exchange otherwise used to import the fertilizer.

15. The 300,000 ton Yangtze ethylene plant in Nanjing

This project produces mainly raw materials for plastics and synthetic fibre. The project involves a total investment of 6.684 billion yuan, including 4.5 billion yuan in foreign loans, to pay for new equipment and modern technology. It is designed to have an annual processing capacity of 3 million tons of crude oil, and produce 300,000 tons of ethylene, 280,000 tons of plastics, 450,000 tons of benzoic acid, 180,000 tons of glycol and 600,000 tons of organic chemical raw materials annually.

16. Steyr Heavy-Duty Automobiles

Construction of the project, involving a total investment of 700 million yuan, began in 1986 and is expected to be completed and put into operation in 1991. It will have an annual capacity of 10,000 heavy-duty motor vehicles and 15,000 engines. The project is also aimed at bringing together the scattered heavy-duty automobile enterprises to form the China Heavy-Duty Automobile Industrial Enterprise Joint Co. By breaking a production pattern characterized by compartmentalization, the joint company will boast an ability to design and produce high-grade heavy-duty automobiles in large batches, thus putting an end to the situation wherein China's heavy-duty automobiles are poor in quality and monotonous in design. The project will help improve the supply of heavy-duty automobiles, now in short supply.

17. Beijing colour kinescope project

It is the largest Sino-Japanese joint venture in the electronics industry. Using a total investment of 500 million yuan, the project will have an annual production capacity of 1.81 million colour picture tubes. Construction of the project started in September 1987, and in July 1989 it began batch-production. It was completed in May 1990, four months ahead of schedule.

18. Second phase of the Yizheng Chemical Fibre project

The Yizheng Chemical Fibre Industrial Joint Co. in Jiangsu is a large enterprise to be built with domestic and foreign funds. The first phase of the project, designed to have an annual capacity of 120,000 tons of polyester fibre and 63,000 tons of polyester chips, has a total investment of 1 billion yuan. It was completed between January 1982 and March 1987. The second phase is designed to have a production capacity of 120,000 tons of polyester fibre and 186,000 tons of polyester chips, using a total investment of 1.5 billion yuan. Construction of the project began in 1985 and is expected to be completed by the end of 1990; production is expected in 1991. The construction of the Yizheng project marks China's entry into a new stage of the chemical fibre industry. It will supply the country with 489,000 tons of chemical fibre and related raw materials, some 25 per cent of China's chemical fibre production capacity.

19. Fushun Detergent Chemical Works in Liaoning

Its designed production capacity includes 50,000 tons of fatty alcohol and 72,000 tons of straight-chain benzol alkyl. Construction of the project began in November 1989 and is expected to be completed by the end of 1992. By then, Fushun will become China's largest detergent raw material production base with an annual capacity of 1 million tons, putting an end to the situation wherein the State chiefly depends on foreign raw materials for synthetic detergent.

20. Hefei State Synchronous Irradiation Laboratory in Anhui

Construction of the Hefei Synchronous Irradiation laboratory, a high-tech project in keeping pace with world advanced level, was planned in 1977 as a State key project and begun in 1984. The project uses a total investment of close to 60 million yuan to construct one 800 million electron volt special electron synchronous irradiation accelerator with a radiant intensity of 100-300 milliampere. The accelerator will be entirely designed, manufactured, installed and trial-operated by China. Thus far, the installation of the major equipment has been completed and the trial-operation, drawing to an end by December 1990, was expected to bear fruit. The project will fill in China's technological blanks and support the effort of those already working with the electron positron collider in China. It will be widely used in physics, chemistry, biology and other basic sciences, as well as in applied and marginal science such as the science of materials, surface science and life research. Its presence will certainly promote the development of China's science and technology.

Source: *Beijing Review* (June 18-24 and June 25-July 1, 1990).

ANNEX D

INDUSTRIAL INVESTMENT INFORMATION AND OPPORTUNITIES

D-1. CHINESE-FOREIGN JOINT VENTURES

A Chinese-foreign joint venture, also known as an equity joint venture, is an enterprise established in China and jointly managed by a foreign firm, enterprise or other economic organization. It is characterized by joint investment and management, with the risks, profits and losses shared by both partners of the joint venture.

According to the *Law of the People's Republic of China on Joint Ventures Using Chinese and Foreign Investment*, partners of a joint venture determine, through discussion, the proportion of investment each of them contributes. The share of the foreign partner(s) should in general be not less than 25 per cent. However, there is no stipulation in respect to the maximum share of the foreign partner(s). Investment may take the form of cash or physical assets (such as factory buildings, premises, facilities, machinery equipment, tools, raw and semi-processed materials, components and parts, warehouses, means of transportation and other means of production), as well as industrial property rights and special technology. Chinese partner(s) may also contribute land-use rights as investment.

In terms of its organizational form, a Chinese-foreign joint venture is a limited liability company, enjoying the status of a legal person subject o the jurisdiction and protection of Chinese laws. The board of directors is the highest organ of power of the joint venture. A joint venture, managed jointly by Chinese and foreign partners, pursues the system of general manager assuming full responsibility under the leadership of the board of directors. Under the terms of the approved contract, the joint venture enjoys the full power of autonomy over its labour management, finance and materials, as well as the production, supply and marketing of its products. A joint venture can, based on its own characteristics, adopt diverse forms of management and operation and run the enterprise in accordance with international practice.

Under Chinese laws, profits are shared according to the proportion of investment contributed by each partner in the registered capital. During the co-operation period, no partner in a joint venture is allowed to withdraw his invested sum of capital. A joint venture is liable for debts only within the limit of all its properties, and risks or losses are shared according to the proportion of investment contributed by each partner who bears the debts only within the limit of the amount of his capital contribution. When disbanded, the joint venture must clear its accounts in accordance with the law.

If the joint venture is an export enterprise or a technologically-advanced enterprise, or is located in a special economic zone, coastal economic development area or open coastal city, it will enjoy more preferential treatment on taxation.

A Chinese-foreign joint venture is required to register with the administrative department for industry and commerce and to apply for a business licence.

Source: *The China Investment Guide*, Fourth Edition, China International Economic Consultants, Inc., 1989.

D-2 CHINESE-FOREIGN CO-OPERATIVE JOINT VENTURES

A Chinese-foreign co-operative joint venture refers to a contractual joint venture. All the liabilities, rights and obligations of the parties, including the investment composition, distribution of benefits as well as the management, are all specified in the contract which is signed through consultations and negotiations among the co-operating parties on the principle of equality and mutual benefit. This kind of venture is characterized by its flexibility, simplicity, and ease of reaching agreement. According to the *Law of the People's Republic of China on Chinese-Foreign Co-operative Joint Ventures*, the co-operating parties can make investment or provide their contributions in the form of cash, physical objects, land-use rights, industrial property, non-patent technology and other proprietary rights. The general practice of forming a Chinese-foreign Co-operative joint venture in China is as follows:

- The foreign party provides the required capital, equipment, material and technology, while the Chinese party provides the land, premises including usable existing equipment and installations, workforce, material resources and a small amount of capital.
- In the contract signed between the co-operative parties, there will be provision for the distribution of benefits, which generally takes the form of the division of profits or the division of products.
- The parties, or one of the parties, can recover their or its investment capital within a certain period of time after the joint venture goes into operation and in certain forms as specified in the contract.
- The fixed assets investment of the foreign party can be repatriated through depreciation after a certain period, which can also be shortened with the approval of the relevant taxation departments.
- The contractual joint venture may form its own Board of Directors or a united management body composed of representatives from the co-operating parties. Within the scope of the approved contract, the joint venture enjoys autonomy in the management which may be performed by the general manager appointed by the Board of Directors or united management body, or may be entrusted to one party or a third party as agreed by all the co-operating parties.
- After termination of the joint venture, the assets settlement procedures vary with different investment and repatriation situations. In general, all assets or part of the assets of a co-operative joint venture shall be turned over to the Chinese party without undergoing any liquidation if the foreign party has already recovered its investment principal in the duration of the joint venture under normal operating conditions.
- According to the *Law on Chinese-Foreign Co-operative Joint Ventures*, the duration of a co-operative joint venture shall be decided by the co-operating parties through consultations, and stipulated in the contract.
- At present, each party pays its income tax separately. The foreign party pays income tax in accordance with the *Income Tax Law of the People's Republic of China Concerning Foreign Enterprises*, and the Chinese party according to Chinese tax law and regulations concerning State enterprises.
- A co-operative joint venture will enjoy much more preferential treatment in taxation if it is an export-oriented enterprise or enterprise with advanced technology, and if it is located in a special economic zone, a costal economic development zone, or in one of the coastal open cities.

Source: *The China Investment Guide*, Fourth Edition, China International Economic Consultants, Inc., 1989.

D-3 WHOLLY FOREIGN-OWNED ENTERPRISE

Wholly foreign-owned enterprises are those established in China by foreign enterprises, other economic organizations or individuals exclusively with their own capital that exercise independent accounting in accordance with the relevant Chinese laws. The term does not include branches set up in China by foreign enterprises or other economic organization.

According to the *Law of the People's Republic of China on Enterprise Operated Exclusively with Foreign Capital*, adopted at the Fourth Session of the Sixth National People's Congress on 12 April 1986, "Enterprises to be established exclusively with foreign capital shall be conducive to the development of China's national economy. Such enterprises shall use advanced technology and equipment or market all or most of their products outside China". These are the fundamental conditions and requirements for setting up wholly foreign-owned enterprises in China.

Before the establishment of a wholly foreign-owned enterprise, the foreign investor should file an application with the Chinese authorities concerned for approval and then register with the departments in charge of administration of industry and commerce and receive a business licence.

A wholly foreign-owned enterprise must be an independent economic entity, setting up accounts books in China and exercising independent accounting.

The wholly foreign-owned enterprises which meet the conditions for being considered a legal person under Chinese laws shall be so considered.

The present wholly foreign-owned enterprises already approved and established in China all meet the conditions for legal persons and have registered in accordance with Chinese laws. They all enjoy the status of a legal person and are subject to the protection of and control by Chinese laws.

A wholly foreign-owned enterprise enjoys full power of autonomy in its operations and management. The enterprise shall be free from interference in its operations and management so long as these are conducted in accordance with the approved articles of association.

A wholly foreign-owned enterprise enjoys full power of autonomy in its operations and management. The enterprise shall be free from interference in its operations and management so long as these are conducted in accordance with the approved articles of association.

A wholly foreign-owned enterprise shall employ Chinese workers and administrative staff under contracts concluded according to law. The enterprise can decide employment and dismissal of workers and staff but the contracts shall include provision relating to employment, dismissal, salaries and wages, welfare, occupational safety and insurance. Workers and staff employed by the wholly foreign-owned enterprise may set up trade unions in accordance with the law, and the enterprise shall provide necessary facilities for the activities of the trade unions.

A wholly foreign-owned enterprise shall pay taxes in accordance with relevant taxation regulations. It may enjoy preferential treatment for reduction of taxes or exemption from them. At present, such enterprises pay their income tax in accordance with the *Income Tax Law Concerning Foreign Enterprises*. Where a wholly foreign-owned enterprise falls into the category of export enterprise or technologically advanced enterprise or is located in a special economic zone or coastal city, it may enjoy more favourable treatment in taxation.

The wholly foreign-owned enterprise should try to balance its foreign exchange receipts and payments. Its products should be exported in return for the foreign exchange needed to pay the lawful profits of investors and wages of its foreign employees. If, with the approval of the competent authorities, the enterprise markets its products in China and consequently experiences an imbalance in foreign exchange, the said authorities should be responsible for eliminating the imbalance.

Wholly foreign-owned enterprises are subject to management and supervision by the departments for the control of foreign exchange, the Customs, accounting, auditing, taxation, industrial and commercial administrations. The Chinese Government reaffirmed in the *Law on Enterprises Operated Exclusively with Foreign Capital* that "The investments made by a foreign investor in China, the profit he earns and his other lawful rights and interests shall be protected by Chinese laws. The wholly foreign-owned enterprise must abide by Chinese laws and statutes and must do nothing detrimental to China's public interests". Except under special circumstances, the State shall not nationalize or expropriate wholly foreign-owned enterprises. Should it prove necessary to do so in the public interest, legal procedures will be followed and reasonable compensation will be made.

Source: *The China Investment Guide*, Fourth Edition, China International Economic Consultants, Inc., 1989.

D-4 PROTECTION FROM DOUBLE TAXATION

Since January 1981, the Chinese Government has started negotiations on taxation with foreign countries and successively signed agreements on the avoidance of double taxation and prevention of evasion have been signed by 20 countries, listed in the Table below. Of these, 18 agreements are now in force. In addition, China has initialled agreements with four countries and is ready to open negotiations with other countries.

	Government	Date of signing	Date of entering into force
1.	Japan	06.09.1983	26.06.1984
2.	France	30.05.1984	20.02.1985
3.	United Kingdom	26.07.1984	23.12.1984
4.	Belgium	18.04.1985	11.09.1987
5.	Germany, Federal Republic of	06.06.1985	04.05.1986
6.	United States of America	30.04.1984	21.11.1986
7.	Malaysia	23.11.1985	14.09.1986
8.	Norway	25.02.1986	21.12.1986
9.	Denmark	26.03.1986	22.10.1986
10.	Singapore	18.04.1986	11.12.1986
11.	Finland	12.05.1986	18.12.1987
12.	Canada	12.05.1986	29.12.1986
13.	Sweden	16.05.1986	03.01.1987
14.	New Zealand	16.09.1986	17.12.1986
15.	Thailand	27.10.1986	29.12.1986
16.	Italy	31.10.1986	..
17.	The Netherlands	13.05.1987	05.03.1987
18.	German Democratic Republic	08.06.1987	14.10.1987
19.	Czechoslovakia	11.06.1987	23.12.1987
20.	Poland	07.06.1988	..

The texts of taxation agreements are drawn up by negotiation with countries which have economic contacts with China, and are worked out according to local conditions, the principle of mutual benefit, and with reference to the model agreements drafted by the United Nations and the Organization of Economic Co-operation and Development. The main points of taxation agreements are as follows:

- the agreements are generally at governmental level. On the Chinese side, an agreement enters into force after it is filed with the State Council, while the other signatory will go through its own legal procedures;

- on the Chinese side, the agreement concerns individual income tax, joint venture income tax, and foreign business income tax. The foreign side should apply the agreement to the same or similar taxes. Other taxes are not included in taxation agreements; and

- China, in principle, adopts international articles such as the definitions of: residents, standing bodies, airlift and sea transportation, joint businesses, dividends, interest rates, fees for using special permits, property revenues, service charges, methods for avoiding double taxation, non-discriminatory treatment, and information exchanges. However, because of different taxation laws in different countries, specific clauses may vary to a certain extent in different agreements.

Source: *The China Investment Guide*, Fourth Edition, China International Economic Consultants, Inc., 1989.

D-5. INCOME TAX LAW OF THE PEOPLE'S REPUBLIC OF CHINA FOR ENTERPRISES WITH FOREIGN INVESTMENT AND FOREIGN ENTERPRISES, JULY 1991

(Adopted at the Fourth Session of the Seventh National People's Congress on April 9, 1991, promulgated by Order No. 45 of the President of the People's Republic of China on April 9, 1991 and effective as of July 1, 1991)

Article 1

Income tax shall be paid in accordance with the provisions of this Law by enterprises with foreign investment within the territory of the People's Republic of China on their income derived from production, business operations and other sources.

Income tax shall be paid in accordance with the provisions of this Law by foreign enterprises on their income derived from production, business operations and other sources within the territory of the People's Republic of China.

Article 2

"Enterprises with foreign investment" referred to in this Law means Chinese-foreign equity joint ventures, Chinese-foreign contractual joint ventures and foreign-capital enterprises that are established in China.

"Foreign enterprises" referred to in this Law means foreign companies, enterprises and other economic organizations which have establishments or places in China and engage in production or business operations, and which, though without establishments or places in China, have income from sources within China.

Article 3

Any enterprise with foreign investment which establishes its head office in China shall pay its income tax on its income derived from sources inside and outside China. Any foreign enterprise shall pay its income tax on its income derived from sources within China.

Article 4

The taxable income of an enterprise with foreign investment and an establishment or a place set up in China to engage in production or business operations by a foreign enterprise, shall be the amount remaining from its gross income in a tax year after the costs, expenses and losses have been deducted.

Article 5

The income tax on enterprises with foreign investment and the income tax which shall be paid by foreign enterprises on the income of their establishments or places set in China to engage in production or business operations shall be computed on the taxable income at the rate of 30 per cent, and local income tax shall be computed on the taxable income at the rate of 3 per cent.

Article 6

The State shall, in accordance with the industrial policies, guide the orientation of foreign investment and encourage the establishment of enterprises with foreign investment which adopt advanced technology and equipment and export all or greater part of their products.

Article 7

The income tax on enterprises with foreign investment established in special economic zones, foreign enterprises which have establishments or places in special economic zones engaged in production or business operations, and on enterprises with foreign investment of a production nature in economic and technological development zones, shall be levied at the reduced rate of 15 per cent.

The income tax on enterprises with foreign investment of a production nature established in coastal economic open zones or in the old urban districts of cities where the special economic zones or the economic and technological development zones are located, shall be levied at the reduced rate of 24 per cent.

The income tax on enterprises with foreign investment in coastal economic open zones, in the old urban districts of cities where the special economic zones or the economic and technological development zones are located or in other regions defined by the State Council, within the scope of energy, communications, harbour, wharf or other projects encouraged by the State, may be levied at the reduced rate of 15 per cent. The specific rules shall be regulated by the State Council.

Article 8

Any enterprise with foreign investment of a production nature scheduled to operate for a period of not less than ten years shall, from the year beginning to make profit, be exempted from income tax in the first and second years and allowed a 50 per cent reduction in the third to fifth years. However, the exemption from or reduction of income tax on enterprises with foreign investment engaged in the exploitation of resources such as petroleum, natural gas, rare metals, and precious metals shall be regulated separately by the State Council. Enterprises with foreign investment which have actually operated for a period of less than ten years shall repay the amount of income tax exempted or reduced already.

The relevant regulations, promulgated by the State Council before the entry into force of this Law, which provide preferential treatment of exemption from or reduction of income tax on enterprises engaged in energy, communications, harbour, wharf and other major projects of a production nature for a period longer than that specified in the preceding paragraph, or which provide preferential treatment of exemption from or reduction of income tax on enterprises engaged in major projects of a non-production nature, shall remain applicable after this Law enters into force.

Any enterprise with foreign investment which is engaged in agriculture, forestry or animal husbandry and any other enterprise with foreign investment which is established in remote underdeveloped areas may, upon approval by the competent department for tax affairs under the State Council of an application filed by the enterprise, be allowed a 15 to 13 per cent reduction of the amount of income tax payable for a period of another ten years following the expiration of the period for tax exemption or reduction as provided for in the preceding two paragraphs.

After this Law enters into force, any modification to the provisions of the preceding three paragraphs of this Article on the exemption from or reduction of income tax on enterprises shall be submitted by the State Council to the Standing Committee of the National People's Congress for decision.

Article 9

The exemption from or reduction of local income tax on any enterprise with foreign investment which operates in an industry or undertakes a project encouraged by the State shall, in accordance with the actual situation be at the discretion of the people's government of the relevant province, autonomous region or municipality directly under the central government.

Article 10

Any foreign investor of an enterprise with foreign investment which reinvests its share of profit obtained from the enterprise directly into that enterprise by increasing its registered capital, or uses the profit as capital investment to establish other enterprises with foreign investment to operate for a period of not less than five years shall, upon approval by the tax authorities of an application filed by the investor, be refunded 40 per cent of the income tax already paid on the reinvested amount. Where regulations of the State Council provide otherwise in respect of preferential treatment, such provisions shall apply. If the investor withdraws its reinvestment before the expiration of a period of five years, it shall repay the refunded tax.

Article 11

Losses incurred in a tax year by any enterprise with foreign investment and by an establishment or a place set up in China by a foreign enterprise to engage in production or business operations may be made up by the income of the following tax year. Should the income of the following tax year be insufficient to make up for the said losses, the balance may be made up by its income of the further subsequent year, and so on, over a period not exceeding five years.

Article 12

Any enterprise with foreign investment shall be allowed, when filing a consolidated income tax return, to deduct from the amount of tax payable the foreign income tax already paid abroad in respect of the income derived from sources outside China. The deductible amount shall, however, not exceed the amount of income tax otherwise payable under this Law in respect of the income derived from sources outside China.

Article 13

The payment or receipt of charges or fees in business transactions between an enterprise with foreign investment or an establishment or a place set up in China by a foreign enterprise to engage in production or business operations, and its associated enterprises, shall be made in the same manner as the payment or receipt of charges or fees in business transactions between independent enterprises. Where the payment or receipt of charges or fees is not made in the same manner as in business transactions between independent enterprises and results in a reduction of the taxable income, the tax authorities shall have the right to make reasonable adjustment.

Article 14

Where an enterprise with foreign investment or an establishment or a place set up in China by a foreign enterprise to engage in production or business operations is established, moves to a new site, merges with another enterprise, breaks up, winds up or makes a change in any of the main entries of registration, it shall present the relevant documents to and go through tax registration or a change or cancellation in registration, with the local tax authorities after the relevant event is registered, or a change or cancellation in registration is made, with the administrative agency for industry and commerce.

Article 15

Income tax on enterprises and local income tax shall be computed on an annual basis and paid in advance in quarterly instalments. Such payments shall be made within 15 days from the end of each quarter and the final settlement shall be made within five months from the end of each tax year. Any excess payment shall be refunded and any deficiency shall be repaid.

Article 16

Any enterprise with foreign investment and any establishment or place set up in China by a foreign enterprise to engage in production or business operations shall file its quarterly provisional income tax return in respect of advance payments with the local tax authorities within the period for each advance payment of tax, and it shall file an annual income tax return together with the final accounting statements within four months from the end of the tax year.

Article 17

Any enterprise with foreign investment and any establishment or place set up in China by a foreign enterprise to engage in production or business operations shall report its financial and accounting systems to the local tax authorities for reference. All accounting records must be complete and accurate, with legitimate vouchers as the basis for entries.

If the financial and accounting bases adopted by an enterprise with foreign investment and an establishment or a place set up in China by a foreign enterprise to engage in production or business operations contradict the relevant tax provisions of the State Council, tax payment shall be computed in accordance with the relevant tax provisions of the State Council.

Article 18

When any enterprise with foreign investment goes into liquidation, and if the balance of its net assets or the balance of its remaining property after deduction of the enterprise's undistributed profit, various funds and liquidation expenses exceeds the enterprise's paid-in capital, the excess portion shall be liquidation income on which income tax shall be paid in accordance with the provisions of this Law.

Article 19

Any foreign enterprise which has no establishment or place in China but derives profit, interest, rental, royalty and other income from sources in China, or though it has an establishment or a place in China, the said income is not effectively connected with such establishment or place, shall pay an income tax of 20 per cent on such income.

For the payment of income tax in accordance with the provisions of the preceding paragraph, the income beneficiary shall be the taxpayer and the payer shall be the withholding agent. The tax shall be withheld from the amount of each payment by the payer. The withholding agent shall, within five days, turn the amount of taxes withheld on each payment over to the State Treasury and submit a withholding income tax return to the local tax authorities.

Income tax shall be exempted or reduced on the following income:

(1) the profit derived by a foreign investor from an enterprise with foreign investment shall be exempted from income tax;

(2) income from interest on loans made to the Chinese Government or Chinese State banks by international financial organizations shall be exempted from income tax;

(3) income from interest on loans made at a preferential interest rate to Chinese State banks by foreign banks shall be exempted from income tax; and

(4) income tax of the royalty received for the supply of technical know-how in scientific research, exploitation of energy resources, development of the communications industries, agricultural, forestry and animal husbandry production, and the development of important technologies may, upon approval by the competent department for tax affairs under the State Council, be levied at the reduced rate of 10 per cent. Where the technology supplies is advanced or the terms are preferential, exemption from income tax may be allowed.

Apart from the aforesaid provisions of this Article, if preferential treatment in respect of reduction of or exemption from income tax on profit, interest, rental, royalty and other income is required, it shall be regulated by the State right to inspect the financial, accounting and tax affairs of enterprises with foreign investment and establishments or places set up in China by foreign enterprises to engage in production or business operations, and have the right to inspect tax withholding of the withholding agent and its payment of the withheld tax into the State Treasury. The entities and the withholding agents being inspected must report the facts and provide relevant information. They may not refuse to report or conceal any facts.

When making an inspection, the tax officials shall produce their identity documents and be responsible for confidentiality.

Article 21

Income tax payable according to this Law shall be computed in terms of Renminbi (Rmb). Income in foreign currency shall be converted into Renminbi according to the exchange rate quoted by the State exchange control authorities for purposes of tax payment.

Article 22

If any taxpayer fails to pay tax within the prescribed time limit, or if the withholding agent fails to turn over the tax withheld within the prescribed time limit, the tax authorities shall, in addition to setting a new time limit for tax payment, impose a surcharge for overdue payment, equal to 0.2 per cent of the overdue tax for each day in arrears, starting from the first day the payment becomes overdue.

Article 23

The tax authorities shall set a new time limit for registration or submission of documents and may impose a fine of 5,000 yuan or less on any taxpayer or withholding agent which fails to go through tax registration or make a change or cancellation in registration with the tax authorities within the prescribed time limit, fails to submit income tax return, final accounting statements, or withholding income tax return to the tax authorities within the prescribed time limit, or fails to report its financial and accounting systems to the tax authorities for reference.

Where the tax authorities have set a new time limit for registration or submission of documents, they shall impose a fine of 10,000 yuan or less on the taxpayer or withholding agent which again fails to meet the time limit for going through registration or making a change in registration with the tax authorities, or for submitting income tax return, final accounting statements or withholding income tax return to the tax authorities. Where the circumstances are serious, the legal representative and the person directly responsible shall be investigated for criminal responsibility, by applying *mutatis mutandis* the provisions of Article 121 of the Criminal Law. This Law does not withhold or withholds an amount less than that should have been withheld, the tax authorities shall set a time limit for the payment of the amount of tax that should have been withheld, and may impose a fine up to but no exceeding 100 per cent of the amount of tax that should have been withheld.

Where the withholding agent fails to turn the tax withheld over to the State Treasury within the prescribed time limit, the tax authorities shall set a time limit for turning over the taxes and may impose a fine of 5,000 yuan or less on the withholding agent; if the withholding agent fails to meet the time limit again, the tax authorities shall pursue the taxes according to law and may impose a fine of 10,000 yuan or less on the withholding agent. If the circumstances are serious, the legal representative and the person directly responsible shall be investigated for criminal responsibility by applying *mutatis mutandis* the provisions of Article 121 of the Criminal Law.

Article 25

Where any person evades tax by deception or concealment or fails to pay tax within the time limit prescribed by this Law and, after the tax authorities pursued the payment of tax, fails again to pay it within the prescribed time limit, the tax authorities shall, in addition to recovering the tax which should have been paid, impose a fine up to but not exceeding 500 per cent of the amount of tax which should have been paid. Where the circumstances are serious, the legal representative and the person directly responsible shall be investigated for criminal responsibility in accordance with the provisions of Article 121 of the Criminal Law.

Article 26

Any enterprise with foreign investment, foreign enterprise or withholding agent, in case of a dispute with the tax authorities on payment of tax, must pay tax according to the relevant regulations first. Thereafter, the taxpayer or withholding agent may, within 60 days from the date of receipt of the tax payment certificate issued by the tax authorities, apply to the tax authorities at the next higher level for reconsideration. The higher tax authorities shall make a decision within 60 days after receipt of the application for reconsideration. If the taxpayer or withholding agent is not satisfied with the decision, it may institute legal proceedings in the people's court within 15 days, from the date of receipt of the notification on decision made after reconsideration.

If the party concerned is not satisfied with the decision on punishment by the tax authorities, it may, within 15 days from the date of receipt of the notification on punishment, apply for reconsideration to the tax authorities at the next higher level than that which made the decision on punishment. Where the party is not satisfied with the decision made after reconsideration, it may institute legal proceedings in the people's court within 15 days from the date of receipt of the decision made after reconsideration. The party concerned may, however, directly institute legal proceedings in the people's court within 15 days from the date of receipt of the notification on punishment. If the party concerned does not apply for reconsideration to the higher tax authorities or institute legal proceedings in the people's court within the time limit, and if the decision on punishment is not fulfilled, the tax authorities which made the decision on punishment may apply to the people's court for compulsory execution.

Article 27

Where any enterprise with foreign investment which was established before the promulgation of this Law would, in accordance with the provisions of this Law, otherwise be subject to higher tax rates or enjoy less preferential treatment of tax exemption or reduction than before the entry into force of this Law, in respect to such enterprise, within its approved period of operation, the law and relevant regulation of the State Council in effect before the entry into force of this Law shall apply. If any such enterprise has no approved period of operation, the law and relevant regulations of the State Council in effect before the entry into force of this Law shall apply within the period prescribed by the State Council. Specific rules shall be regulated by the State Council.

Article 28

Where the provisions of the tax agreement concluded between the Government of the People's Republic of China and foreign governments are different from the provisions of this Law, the provisions of the respective agreement shall apply.

Article 29

Rules for implementation shall be formulated by the State Council in accordance with this Law.

Article 30

This Law shall enter into force on July 1, 1991, the Income Tax Law of the People's Republic of China for Chinese-Foreign Equity Joint Ventures and the Income Tax Law of the People's Republic of China for Foreign Enterprises shall be annulled as of the same date.

Appendix:

The relevant articles in the Criminal Law

Article 121 in case of tax evasion or refusal to pay taxes in violation of tax laws and regulations if the circumstances are serious, the taxpayer shall be ordered to pay the tax due and may be fined in accordance with the tax laws and regulations; the person directly responsible shall also be sentenced to fixed-term imprisonment of not more than three years or criminal detention.

Source: Ministry of Foreign Relations and Trade.

D-6. FOREIGN INVESTMENT DATA

Annex Table D-6.1. Sector-wise distribution of foreign investment, 1990
($10,000)

Industry	Total: Projects		Total: Investment		Direct investment of foreigner: Projects		Direct investment of foreigner: Investment	
	Number	Percent in total	Value	Percent in total investment	Number	Percent in total	Value	Percent in total investment
Total	**7371**	**100.0**	**1208569**	**100.0**	**7273**	**100.0**	**659611**	**100.00**
1.Farming, forestry, animal husbandry, fishery and water conservancy	230	3.1	92874	7.7	223	3.1	12225	1.90
2.Industry	6672	90.5	769912	63.7	6591	90.6	556918	84.40
3.Geological survey and prospecting	1	..	40	..	1	..	40	..
4.Construction	43	0.6	18108	1.5	43	0.6	18108	2.70
5.Transportation, postal and telecommunication services	57	0.8	38331	3.2	51	0.7	3646	0.60
6.Commerce, food service, material supply and marketing and storage	96	1.3	10935	0.9	96	1.3	10660	1.60
7.Real estate management, public, residential and consultancy services	158	2.2	48041	4.0	158	2.2	45247	6.90
8.Health care, sports and social welfare	15	0.2	3798	0.3	15	0.2	3798	0.60
9.Education, culture, art, radio and television broadcasting	16	0.2	5506	0.4	15	0.2	506	
10.Scientific research, polytechnical services	24	0.3	3195	0.3	24	0.3	3195	0.50
11.Banking and insurance								
12.Government agencies, parties and organizations	59	0.8	217829	18.0	56	0.8	5268	0.80
13.Others								

Source: State Statistical Bureau of China.

Annex Table D-6.2. Foreign capital actually used through contracts or agreements, 1990 ($10,000)

Country (region)	Total	Foreign loans	Direct investment by foreign businessmen	Other
United States	53396	16828	35782	786
United Kingdom	20169	8046	11903	220
Germany	13304	8570	4564	170
France	12875	11603	1244	28
Belgium	2907	2301	606	
Sweden	17429	17084	345	
Switzerland	5560	2781	1855	924
Canada	27105	25549	1495	61
Japan	165488	117642	45700	2146
Hong Kong and Macao	450134	24367	394349	31418
Australia	4410	2627	1744	39
New Zealand	89		89	
Total	**1208569**	**509937**	**659611**	**39021**

Source: State Statistical Bureau of China.

Annex Table D-6.3. Categories of direct foreign investment in Xiamen SEZ, 1989 and 1990 ($10,000)

	Number of contracts newly signed		Total value of foreign capital through signed contracts or agreements		Total value of foreign capital actually utilized	
	1989	1990	1989	1990	1989	1990
Total	**201**	**248**	**51449**	**38118**	**12980**	**5273**
1. Farming, forestry, animal husbandry, fishery and water conservancy	9	2	758	811		
2. Industry	172	226	40958	30446		
3. Geological survey and prospecting						
4. Construction		1		54		
5. Transportation, postal and telecommunication services	2	1	131	21		
6. Commerce, food service, material supply and marketing and storage	2	2	56	93		
7. Real estate management, public, residential and consultancy services	15	14	9449	6163		
8. Health care, sports and social welfare						
9. Education, culture, art, radio and television broadcasting						
10. Scientific research, polytechnical services						
11. Banking and insurance						
12. Government agencies, parties and organizations						
13. Others	1	2	97	530		

Source: State Statistical Bureau of China.

Annex Table D-6.4. Categories of direct foreign investment in Hainan Province, 1989 and 1990 ($10,000)

	Number of contracts newly signed		Total value of foreign capital through signed contracts or agreements		Total value of foreign capital actually utilized	
	1989	1990	1989	1990	1989	1990
Total	**378**	**252**	**28060**	**12882**	**10943**	**10055**
1. Farming, forestry, animal husbandry, fishery and water conservancy	27	29	1864	804	672	568
2. Industry	217	155	17648	5562	4214	3270
3. Geological survey and prospecting						
4. Construction	25	20	638	1668	711	414
5. Transportation, postal and telecommunication services	11	4	1491	44	202	244
6. Commerce, food service, material supply and marketing and storage	23	13	579	398	914	531
7. Real estate management, public, residential and consultancy services	66	27	5112	4170	1991	3994
8. Health care, sports and social welfare						
9. Education, culture, art, radio and television broadcasting	2		29		4	
10. Scientific research, polytechnical services	4		101		10	
11. Banking and insurance					53	
12. Government agencies, parties and organizations						
13. Others	3	4	598	286	2172	1034

Source: State Statistical Bureau of China.

Annex Table D-6.5. Categories of direct foreign investment in Shenzhen SEZ, 1989 and 1990 ($10,000)

	Number of contracts newly signed		Total value of foreign capital through signed contracts or agreements		Total value of foreign capital actually utilized	
	1989	1990	1989	1990	1989	1990
Total	**556**	**563**	**37992**	**50156**	**27144**	**34920**
1. Farming, forestry, animal husbandry, fishery and water conservancy	3	2	69	35	140	
2. Industry	519	541	28624	38649	19547	28045
3. Geological survey and prospecting						
4. Construction		2		138		6
5. Transportation, postal and telecommunication services	3	3	643	378	53	195
6. Commerce, food service, material supply and marketing and storage	12	6	3499	5197	1071	815
7. Real estate management, public, residential and consultancy services	9	2	3915	3022	1530	5576
8. Health care, sports and social welfare						
9. Education, culture, art, radio and television broadcasting	4		300		4664	
10. Scientific research, polytechnical services	4		108		33	
11. Banking and insurance						
12. Government agencies, parties and organizations						
13. Others	2	7	834	2737	106	283

Source: State Statistical Bureau of China.

Annex Table D-6.6. Categories of direct foreign investment in Zhuhai SEZ, 1989 and 1990 ($10,000)

	Number of contracts newly signed		Total value of foreign capital through signed contracts or agreements		Total value of foreign capital actually utilized	
	1989	1990	1989	1990	1989	1990
Total	**192**	**347**	**9416**	**23793**	**4296**	**6219**
1. Farming, forestry, animal husbandry, fishery and water conservancy	3	2	124	108	119	81
2. Industry	179	329	7755	15603	3916	5515
3. Geological survey and prospecting						
4. Construction	2	1	19	..	118	6
5. Transportation, postal and telecommunication services	2		151		10	
6. Commerce, food service, material supply and marketing and storage	2	5	214	963	22	41
7. Real estate management, public, residential and consultancy services	2	9	1120	7194	90	485
8. Health care, sports and social welfare						
9. Education, culture, art, radio and television broadcasting	2		33		21	
10. Scientific research, polytechnical services						
11. Banking and insurance						
12. Government agencies, parties and organizations						
13. Others		1		21		91

Source: State Statistical Bureau of China.

Annex Table D-6.7. Categories of direct foreign investment in Shantou SEZ, 1989 and 1990 ($10,000)

	Number of contracts newly signed		Total value of foreign capital through signed contracts or agreements		Total value of foreign capital actually utilized	
	1989	1990	1989	1990	1989	1990
Total	**138**	**141**	**12089**	**14768**	**5891**	**6125**
1. Farming, forestry, animal husbandry, fishery and water conservancy	2	2	261	42	385	34
2. Industry	125	133	9699	13067	4410	4855
3. Geological survey and prospecting						
4. Construction						
5. Transportation, postal and telecommunication services					333	
6. Commerce, food service, material supply and marketing and storage	1		51		3	
7. Real estate management, public, residential and consultancy services	10	6	2078	1659	895	1236
8. Health care, sports and social welfare						
9. Education, culture, art, radio and television broadcasting					13	
10. Scientific research, polytechnical services						
11. Banking and insurance						
12. Government agencies, parties and organizations						
13. Others					122	

Source: State Statistical Bureau of China.

Annex Table D-6.8. Actual utilization of direct foreign investment in four SEZs, 1989 and 1990

	Total of four SEZ		Shenzhen		Zhuhai		Shantou		Xiamen	
	1989	1990	1989	1990	1989	1990	1989	1990	1989	1990
Number of contracts newly signed	1087	1299	556	563	192	347	138	141	201	248
Total value of foreign capital through signed contracts or agreement ($10,000)	110946	126835	37992	50156	9416	23793	12089	14768	51449	38118
Total value of foreign capital actually utilized ($10,000)	50311	52627	27144	34920	4296	6219	5891	6215	12980	5273

Source: State Statistical Bureau of China.

ANNEX E

INSTITUTIONS SUPPORTING INDUSTRY

E-1. Main international trust and investment corporations

China International Trust and Investment Corporation (Holdings)

Address:	CITIC Building, 19 Jianguomenwai Dajie, Beijing
Telephone:	SB: 5002255
Telex:	22305 CITIC CN
Fax:	(01) 5001535
Cable:	3319 Beijing
P.O.Box:	9405 Beijing, China

Organizations abroad:

CITIC Corporation Hong Kong (Holding) Ltd

Address:	25th floor, Admiralty Centre, Tower 1, 18 Harcourt Rd, Hong Kong
Telephone:	5-8410838
Telex:	62685 CITIH HX
Fax:	5-8651625
Cable:	CITIC HX

CITIC Representative Office in Japan

Address:	3rd floor, the LANDIC Third Akasaka Bldg, 2-3-2 Akasaka, Minato-ku, Tokyo, Japan
Telephone:	(03) 5842635 (3) 5842636
Telex:	2424392 CITIC RJ
Fax:	(3) 5056235

CITIC Representative Office in New Yor

Address:	Second World Trade Center, Suite 2250, New York, NY 10048, United States of America
Telephone:	(212) 938-0416
Telex:	229281 CITIC UR
Fax:	(212) 4881107

CITIC Representative Office in Europe

Address:	Bockenheimer Landstrasse 51-53, D-6000 Frankfurt/Main, Germany
Telephone:	(069) 728607-8
Telex:	4170740 CTIC D
Fax:	(069) 728319

CITIC Representative Office in Paris

Address:	44 bis, Rue Pasquier 75008 Paris, France
Telephone:	42935160
Telex:	281424F CITIC PA
Fax:	42949847

The Ka Wah Bank Ltd

Address:	232 Des Voeux Rd, Central, Hong Kong
Telephone:	(852) 5-457131-8
Telex:	74636 KWBNK HX
Fax:	(852) 5-417029

CITIFOR Inc.

Address:	4270 First Interstate Center, 999 Third Avenue, Seattle Wa 98104, United States of America
Telephone:	(206) 6223770
Telex:	0234974147 CIFOUI
Fax:	(206) 6226714

CITIC Canada Ltd

Address:	1870-401 West Georgia St, Box 139 Vancouver, BC, Canada V6B 5A1
Telephone:	(604) 6817204
Telex:	(045) 08719
Fax:	(604) 6817230

CITIC Australia Ltd

Address:	43rd florr, ANZ Tower, 55 Collins St, Melbourne, Vic 3000, Australia
Telephone:	(03) 6547500
Telex:	AA 152920 CITIC A
Fax:	(03) 6547614

Tianjin International Trust and Investment Corporation

Address:	35 Guilin Rd, Heping District, Tianjin
Telephone:	317147, 317143
Telex:	234021 or 23255 TITIC CN

Guangdong International Trust and Investment Corporation

Address:	11th floor, Garden Building, Garden Hotel 368 Huan Shi Rd, Guangzhou, Guangdong
Telephone:	338999 ext. 71101, 71157
Telex:	44422 GITIC CN
Fax:	346890
Cable:	0975 Guangzhou

Shanghai Investment and Trust Corporation

Address:	3rd floor, Lianyidaxia, 100 Yanan East Rd, Shanghai
Telephone:	284120
Telex:	33031 SITCO CN
Cable:	Investco Shanghai
P.O.Box:	252 Shanghai

Office in San Francisco, United States of America

Address:	909 Montgomery St, San Francisco, California, United States of America
Telephone:	(415) 3920916
Telex:	3716938 SITCO SFO

Anhui International Trust and Investment Corporation

Address:	29 Huizhou Rd, Hefei, Anhui
Telephone:	76670, 73468, 56421, 65403, 72278
Telex:	90006 AHINT CN
Cable:	1386 Hefei

Hunan International Trust and Investment Corporation

Address:	43 Shaoshan Rd, Changsha, Hunan Province
Telephone:	53960, 52032
Telex:	98110 HITIC CN
Cable:	0033 Changsha

Hubei International Trust and Investment Corporation

Address:	2 Nanjing Rd, Wuhan
Telephone:	23409
Telex:	40110 HBEXT CN
Fax:	512857

China Venturetech Investment Corporation

Address:	5th floor, Huizhong Hotel, 120 Zhushikou St (W), Xuanwu District, Beijing
Telephone:	3015320, 3015318, 3014810
Telex:	222506
Fax:	3013965
P.O.Box:	421 Beijing, China

Source: China International Economic Consultants, Inc and CITIC Publishing House, *The China Investment Guide*, Fourth Edition, Longman Group Limited (Hong Kong 1989).

E-2. Main economic consulting corporations

China International Economic Consultants, Inc.

Address:	2nd floor, CITIC Bldg, 19 Jianguomenwai Dajie, Beijing
Telephone:	5002255 ext. 3261, 5003422
Telex:	22994 CIEC CN
Fax:	(01) 5003721

China Investment Consulting Corporation

Address:	9 Fuxing Rd, Beijing
Telephone:	8013834, 8011050, 862238
Telex:	222617 CICC
Cable:	5189 Beijing
P.O.Box:	3821 Beijing, China

China International Engineering Consulting Corporation

Address:	Jianweidalou, Baiwanzhuang, Western Suburb, Beijing
Telephone:	8022266, 895561, 866143
Telex:	22095 CIECC
Fax:	(861) 8013985
Cable:	6696 Beijing
P.O.Box:	850 Beijing, China

Member Consultancy Corporations

Iron and Steel and Non-Ferrous Metals Industry:

China Guangan Engineering Consulting Corporation
China Jingxi Engineering Consulting Corporation
China Jiangxia Engineering Consulting Corporation
China Southern Engineering Consulting Corporation
Jiangnan Engineering Consulting Corporation
Beijing Zhongye Building Consulting Corporation
Beijing International Engineering Consulting Corporation
Anshan Qianshan Engineering Consulting Corporation
Changsha Hongda Engineering Consulting Corporation
Zhongyuan Engineering Consulting Corporation

Chemical and Petrochemical Industry:

China Chemical Industry Consultancy Service Corporation
Beijing Chemical Engineering Consulting Corporation
Chengdu Chemical Engineering Consulting Corporation
China Petrochemical Industry Consulting Corporation
Beijing Petroleum Chemical Engineering Consulting Corporation
Shanghai Oriental Engineering Consulting Corporation

Building Materials Industry:

China Building Materials Engineering Consulting Corporation
Tianjin Cement Industry Engineering Consultancy Corporation

Water Conservancy and Electric Power Industry:

China Water Conservancy and Hydroelectric Power Engineering Consulting Corporation
China Power Engineering Consulting Corporation
Beijing Desheng Power Engineering Consulting Corporation

Coal Industry:

China National Coal Development Consulting Corporation
Technical Consultancy Committee of the Ministry of Coal Industry

Railways, Transportation and Civil Aviation:

China Roads Engineering Consulting Corporation
China Subway Engineering Design Consulting Corporation
China Hua Yang Engineering Consulting Corporation
China Railway Science Technology Consulting Corporation
China Shipping Engineering Consulting Corporation
Shanghai Donggan Shipping Engineering Consulting Corporation
China Civil Aviation Engineering Consulting Corporation

Machinery Industry:

China Machinery Engineering Consulting Corporation
Xinghua Engineering Consulting Corporation
Beijing Yinyan Engineering Consulting Corporation
China Wuzhou Engineering Consulting Corporation
Shanghai Zhenhua Engineering Consulting Corporation

Electronics Industry:

Beijing Consulting Corporation for Electronic Engineering
Sichuan Consulting Corporation for Electronic Engineering
China Souther Electronic System Engineering Corporation

Textiles and Light Industry:

China Textile Engineering Consulting Corporation
China United Light Industrial Engineering Consulting Corporation
Beijing Engineering Consulting Corporation of Light Industry
Shanghai Engineering Consulting Corporation of Light Industry
Guangzhou Engineering Consulting Corporation of Light Industry
Changsha Engineering Consulting Corporation of Light Industry

Building Industry:

Northwest China Building Engineering Consulting Corporation
Southwest China Building Engineering Consulting Corporation
Northeast China Building Engineering Consulting Corporation
China Building Design Consulting Corporation
China Urban Planning Design Consulting Corporation
China Refrigeration and Air-Conditioning Engineering Consulting Corporation

Forestry Industry:

China Forestry Industry Engineering Consulting Corporation

Comprehensive-nature Industry:

Beijing Engineering Consulting Corporation
Shanxi Engineering Consulting Corporation
Shanghai Investment Consulting Corporation
Zhejiang Economic Construction Planning Institute
Fujian Engineering Consulting Corporation
Hunan Engineering Consulting Corporation
Liaoning National Defence Engineering Consulting Corporation

Source: China International Economic Consultants, Inc and CITIC Publishing House, *The China Investment Guide*, Fourth Edition, Longman Group Limited (Hong Kong 1989).

E-3. Financial institutions permitted to act as guarantors in dealing in foreign exchange business

Banks:
The Bank of China and its branches
The Shanghai Branch of the Hong Kong and Shanghai Banking Corporation
The Shanghai Branch of Standard Chartered Bank
The Shanghai Branch of Overseas Chinese Banking Corporation Ltd
The Shanghai Branch of the Bank of East Asia Ltd
The Shenzhen Branch of Nanyang Commercial Bank
The Shekou Branch of Nanyang Commercial Bank
The Zhuhai Branch of Nantong Bank
The Shenzhen Branch of Guangdong Provincial Bank
The Investment Bank of China
The Xiamen International Bank
The Shenzhen Branch of the Hong Kong and Shanghai Banking Corporation
The Shenzhen Branch of Credit and Commerce International (Overseas) Ltd
The Xiamen Branch of United Overseas Bank
The Shenzhen Branch of Banque Indosuez
The Shenzhen Branch of Standard Chartered Bank
The Xiamen Branch of Chiyu Banking Corporation Ltd
The Shenzhen Branch of Société Générale
The Xiamen Branch of the Hong Kong and Shanghai Banking Corporation
CITIC Industrial Bank
People's Construction Bank of China

ITICs:
China International Trust and Investment Corporation
Shanghai Investment and Trust Corporation
Guangdong International Trust and Investment Corporation
Fujian Investment and Enterprise Corporation
Hubei International Trust and Investment Corporation
Tianjin International Trust and Investment Corporation
Zhejiang International Trust and Investment Corporation
The Bank of China Trust and Consultancy Company
Liaoning International Trust and Investment Corporation
Hebei International Trust and Investment Corporation
Guangzhou International Trust and Investment Corporation
Jiangsu International Trust and Investment Corporation
Beijing International Trust and Investment Corporation
Jiangxi International Trust and Investment Corporation
Shenzhen International Trust and Investment General Corporation
Shenyang International Trust and Investment Corporation
Chongqing International Trust and Investment Corporation
Shaanxi Joint Financial Investment Corporation
The Shenzhen Branch of Guangdong International Trust and Investment Corporation
The Commercial Bank of China Trust and Investment Corporation
Zhongyuan Development Trust and Investment General Corporation (Henan)
Sichuan Changjiang International Trust and Investment Corporation
Anhui International Trust and Investment Service Corporation
Jilin International Trust and Investment Corporation
Ningxia Islam International Trust and Investment Corporation
Shantou International Trust and Investment Corporation
Henan International Trust and Investment Corporation

Shanxi Economic Development and Investment Corporation
Guangxi International Trust and Investment Corporation
Hunan International Trust and Investment Corporation
China Venturetech Investment Corporation
Dalian International Trust and Investment Corporation
Nanjing International Trust and Investment Corporation
Lianyungang International Trust and Investment Corporation
Xiamen International Trust and Investment Corporation
Shandong International Trust and Investment Corporation
China Agribusiness Trust and Investment Corporation

Source: China International Economic Consultants, Inc and CITIC Publishing House, *The China Investment Guide*, Fourth Edition, Longman Group Limited (Hong Kong 1989).

ANNEX F

MAIN INDICATORS OF THE TOP 100 INDUSTRIAL ENTERPRISES IN CHINA, 1990

Annex F. **Main indicators of the top 100 industrial enterprises in China, 1990**
(Ranked by sales income in Rmb)

Enterprises	Sales income	Gross industrial output Value (at current prices)	Original value of fixed assets	Net value of fixed assets	Total profits and taxes	Total workers and staff at year-end
1. Huadong Power Union Corporation	11064820	10129080	21373940	15010490	2500400	144327
2. Daqing Petroleum Administration Bureau	10215506	17087370	18269356	13245113	2208983	128730
3. Huabei Power Union Corporation	9031530	8351280	21001650	14469880	2417070	14596101
4. Huazhong Power Service Administration Bureau	8498390	7599520	18945760	13704910	2103230	152448
5. Dongbei Power General Corporation	8391060	8460760	22928530	15812420	1098680	144068
6. Anshan Iron and Steel Corporation	7744650	8260770	13781780	8648740	1823140	217303
7. Capital Iron and Steel Corporation	6902998	6160550	4862248	3207932	2517903	134516
8. Wuhan Iron and Steel Corporation	6633880	6170050	7428960	4702180	1824060	128899
9. Shanghai Petroleum and Chemical Industry General Works	6234292	120978	4826930	3137630	1299989	59976
10. Beijing Yanshan Petroleum and Chemical Industry Corporation	5809189	4892590	4284833	2348511	1769478	39476
11. Shengli Petroleum Administration Bureau	5399594	7631466	19257962	15367699	-191527	98666
12. Qilu Petroleum and Chemical Industry Corporation	5121925	5475390	7110409	6130746	1193047	46671
13. Shanghai Baoshan Iron and Steel General Works	4817766	160337	13401471	11797046	1037498	31916
14. Shandong Power Industry Bureau	4495147	4017610	6928637	5105051	1037503	45152
15. Guangdong Power Industry Bureau	4382889	2799228	8221468	6717441	851121	32847
16. Jilin Chemical Industry Corporation	4282854	4353480	4582716	3035719	1128324	71042
17. Xibei Power Service Administration Bureau	4173170	4051610	14104530	10599550	1390230	86506
18. Daqing Petroleum and Chemical Industry General Works	4067883	4227190	6308775	4838372	1138462	35953
19. No.2 Motor Vehicle Plant	3874507	4315428	3152570	1903451	585884	79075
20. Fushun Petroleum and Chemical Industry Corporation	3775988	3679160	2435837	1511256	838545	37437
21. Maoming Petroleum Industry Corporation	3306179	3304970	1790211	980396	650404	26396
22. Yizheng Chemical Fibre Industry Union Corporation	3270591	3200777	1301930	1104256	566106	19028
23. Jinling Petroleum and Chemical Industry Corporation	3205087	3257800	2215094	1479178	547124	24598
24. Yangzi Petroleum and Chemical Industry Corporation	3154106	3149540	6904975	6480800	748751	19823
25. Liaohe Petroleum Prospecting Bureau	2982782	3515648	10639304	7365547	-339636	65950
26. Shanghai Gaoqiao Petroleum and Chemical Industry Corporation	2916600	227780	1743137	1292161	607615	19050
27. Yuxi Cigarette-Making Factory	2885541	2885020	373601	308736	1884002	4770
28. Benxi Iron and Steel Corporation	2665671	2784790	4481881	3054940	590928	72132
29. Shanxi Taiyuan Iron and Steel Corporation	2659606	2859432	2702357	1551958	585107	62198
30. No.1 Motor Vehicle Plant	2632941	3633001	2028784	917125	215410	73628

(continued)

Annex F. (continued)

Enterprises	Sales income	Gross industrial output Value (at current prices)	Original value of fixed assets	Net value of fixed assets	Total profits and taxes	Total workers and staff at year-end
31. Sichuan Power Industry Bureau	2580072	2580070	6133055	4030358	864979	58149
32. Kunming Cigarette-Making Factory	2505723	2522629	294756	257733	1747169	4033
33. Ma'anshan Iron and Steel Corporation	2451581	2392574	2778785	1804308	535275	62649
34. Baotou Iron and Steel Corporation	2449193	2227660	3705632	2511579	444742	77340
35. Baling Petroleum and Chemical Industry Corporation	2430830	2414940	2268405	1545302	383309	31537
36. Shanghai Cigarette-Making Factory	2359708	30095	267481	233415	1476561	5574
37. Tianjin Petroleum and Chemical Industry Corporation	2333870	2005470	2366260	1569020	523580	23473
38. Dalian Petroleum and Chemical Industry Corporation	2102677	2125054	1143731	770716	608226	11280
39. Shanghai No.1 Iron and Steel Plant	2048072	25499	762611	490743	11099	22448
40. Shanghai No.3 Iron and Steel Plant	1962070	119937	887434	622507	179345	22259
41. Lanzhou Oil Refinery	1894413	1824844	1003584	530138	320440	16028
42. Guangzhou Petroleum and Chemical Industry General Works	1843939	1845580	983279	686141	189292	10444
43. Shanghai Dazhong Motor Vehicle Limited Company	1822966	2610	674467	592850	268137	2964
44. Liaoyang Petroleum and Chemical Fibre Corporation	1760090	1728350	3112720	2079430	497860	30148
45. Tangshan Iron and Steel Corporation	1750100	1937900	1230860	948530	174910	50278
46. Anqing Petroleum and Chemical Industry General Works	1743786	1591520	972285	628795	234468	13173
47. Chongqing Iron and Steel Corporation	1674251	1873725	1809921	1326079	13693	50391
48. Shanghai No.5 Iron and Steel Plant	1643231	1762	698175	409528	159832	23161
49. Guiyang Cigarette-Making Factory	1627831	1698234	185400	154250	986016	6572
50. Panzihua Iron and Steel Corporation	1583390	1715320	2269458	1565207	181561	61585
51. Beijing Jeeps and Trucks Limited Company	1582807	1578236	293059	198783	210841	4923
52. Jinxi Refining Oil and Chemical Industry General Works	1569776	1577980	779949	468266	286770	10941
53. Jiangxi Copper Corporation	1540689	1491780	2054662	1525025	218639	29108
54. Zhenhai Petroleum and Chemical Industry General Works	1476441	1520280	1045572	759441	303225	8854
55. Sichuan Petroleum Administration Bureau	1420353	1389243	4128705	1966051	23841	51614
56. Xinjiang Petroleum Administration Bureau	1419558	3733335	8937589	6435188	15729	60037
57. Zhongyuan Petroleum Prospecting Bureau	1416685	1891412	9029153	7719227	-379613	42345
58. Wanbao Electrical Equipment Union Corporation	1414540	1594130	803860	573480	36177	18521
59. Hegang Mining Administration Bureau	1410495	1144698	2559942	1675692	-155050	102068
60. Datong Mining Administration Bureau	1337015	1949088	4484500	3045532	-195560	119820
61. Shanghai Papermaking Corporation	1329637	1338497	446076	289345	200272	20169
62. Pingdingshan Mining Administration Bureau	1313169	963122	2488178	1555115	63397	94363

(continued)

Annex F. (continued)

Enterprises	Sales income	Gross industrial output Value (at current prices)	Original value of fixed assets	Net value of fixed assets	Total profits and taxes	Total workers and staff at year-end
63. Jinzhou Petroleum and Chemical Industry Corporation	1272549	1299530	609335	309465	320632	16362
64. Jinan Iron and Steel Plant	1261335	1290696	999059	713948	183045	28470
65. Jingmen Oil Refinery	1255427	1259230	758175	350306	280633	11623
66. Fujian Power Industry Bureau	1228368	637280	2506973	1906043	241580	14546
67. Henan Anyang Iron and Steel Corporation	1216958	1195860	810529	541445	282820	30505
68. No.1 Tractor Factory	1191717	1047135	862416	389921	143147	38206
69. Tianjin Steel Works	1168480	1320710	449600	284570	2320	15277
70. Changde Cigarette-Making Factory	1168392	1199300	209996	180554	674775	4755
71. Shanghai No.1 TV Factory	1161361	28187	120391	83278	124958	3726
72. Fushun Mining Administration Bureau	1159515	1027388	2115443	1134726	15338	81700
73. Lanzhou Chemical Industry Corporation	1151203	1184263	1752702	1038131	227213	28714
74. State-Run Changhong Machinery Factory	1150881	1228289	167641	114548	200919	4823
75. Shanghai No.2 Iron and Steel Plant	1149826	17013	274638	206804	145837	7065
76. Jinbei Motor Vehicle Limited-Liability Company	1142056	1060520	692975	543190	-112471	51124
77. Huabei Petroleum Administration Bureau	1129263	1334454	5590633	3301681	-675513	42924
78. Jixi Mining Administration Bureau	1115937	722413	2321777	1457947	-221800	98339
79. Chengdu Rolling Steel Pipe Factory	1111350	1193160	839290	494330	220550	25342
80. Jinchuan Non-Ferrous Metal Company	1102353	1113174	1460933	1056849	428651	31350
81. Handan Iron and Steel General Works	1094079	1249060	769009	523875	188023	27032
82. Beijing Car and Motor Union Corporation	1093957	572268	197945	120474	68487	9561
83. Nanjing Motor Vehicle Plant	1091472	1026206	466232	284881	59037	20120
84. Great Wall Special Steel Corporation	1082209	1288858	845299	590275	255803	26706
85. Quqing Cigarette-Making Factory	1066139	1067220	156502	129843	695710	3263
86. Wuhan Petroleum and Chemical Industry Works	1047338	978673	376815	265898	157858	5328
87. Huaibei Mining Administration Bureau	1021394	761721	2872468	2259453	-45203	108854
88. Shanghai Electric Wire and Cable Union Corporation	1010999	1072075	207461	154762	139769	8625
89. Shengyang Smeltery	999284	1531910	312684	114565	68547	9374
90. Shanghai Non-Ferros Metal General Corporation	975905	1368424	466746	301315	119475	19080
91. Huludao Zinc Plant	975380	930680	500280	327680	157300	9019
92. Kailuan Mining Administration Bureau	974730	812820	3924700	2808450	-260850	125471
93. Huainan Mining Administration Bureau	971536	854400	2377533	1834598	-134079	107916
94. Tonghua Iron and Steel Corporation	969764	926952	862056	627085	151502	28127

(continued)

Annex F. (continued)

Enterprises	Sales income	Gross industrial output Value (at current prices)	Original value of fixed assets	Net value of fixed assets	Total profits and taxes	Total workers and staff at year-end
95. Inner Mongolia Power Service Administration Bureau	964698	17680	23233	19210	264505	618
96. Shaanxi Colour Trichromoscope Factory	962946	921213	776739	467292	154833	9996
97. Baiyin Non-Ferrous Metal Company	954440	1107099	1166465	678022	52304	38986
98. State-Run Nanjing Radio Works	952339	1256560	170602	95743	93437	7149
99. Jiujiang Oil Refinery	952325	1009300	343603	243829	150862	5152
100. Luoyang Oil Refinery	947614	1206740	571678	482758	247520	6981

Source: State Statistical Bureau of China.

ANNEX G

APPROVED AND/OR OPERATIONAL TECHNICAL CO-OPERATION PROJECTS OF UNIDO

Annex G. Approved and or/operational technical co-operation projects of UNIDO

Project Number	Backstopping responsibility	All.Acc. Code	Project title
DP/CPR/84/016*	IO/IIS/INFR	J12101	Assistance to Shanghai Research Institute (SRIM) of materials on non-destructive testing (phase II) (continuation of DP/CPR/80/052)
DP/CPR/85/010**	IO/IIS/INFR	J12102	Assistance in development and application of precise DC and AC electric quantity measurement techniques
DP/CPR/85/025**	IO/IIS/INFR	J12102	Development Centre for Dimensional Measuring Instruments and Machines
DG/CPR/87/027*	IO/IIS/IMR	J12206	Integrated enterprise management system in Beijing No. 1 Machine Tool Plant (BYJC)
DP/CPR/85/055**	IO/T/AGRO	J13102	The China Garment Technology Development Centre
DG/CPR/85/057*	IO/T/AGRO	J13102	Strengthening of the China Ramie Technology Development Centre
DG/CPR/87/017	IO/T/AGRO	J13102	Strengthening of the China Dyeing and Finishing Development Centre (continuation of DP/CPR/85/056)
DP/CPR/85/023**	IO/T/AGRO	J13103	Assistance of the Beijing Food Additives Development Centre
US.CPR/90/277	IO/T/AGRO	J13104	Leather and leather products journal 'Progress' (see also US/CPR/83/130)
SI/CPR/90/801	IO/T/AGRO	J13104	Assistance to Shanghai Leather Corporation in effluent treatment
DP/CPR/85/006*	IO/T/MET	J13208	Technical training and consultancy service of metallic corrosion and protection
DP/CPR/85/007**	IO/T/MET	J13210	Development of brazing alloy and materials for lead frame in the electronic industry
DP/CPR/85/014*	IO/T/MET	J13210	Research and development of the production technology of tungsten wire for lamp making
DP/CPR/85/043*	IO/T/MET	J13210	Establishment of a Heat Treatment Centre in the Shanghai Institute for Machine Building Technology
XP/CPR/90/005*	IO/T/MET	J13211	Technological testing and techno-economic feasibility study of processing titanium sponge to titanium products by powder metallurgy
DG/CPR/85/018*	IO/T/ENG	J13312	Upgrading of economy and reliability of the locomotive Diesel engine
DG/CPR/91/322*	IO/T/ENG	J13312	Improvement of the foundry machinery industry in China (in co-operation with IO/T/MET)
DG/CPR/91/323*	IO/T/ENG	J13312	Establishment of a pilot autonomous manufacturing island
DG/CPR/91/324*	IO/T/ENG	J13312	Development of turrets for NC lathes
DG/CPR/91/325*	IO/T/ENG	J13312	Modular design of a family of small- to medium-size machining centres
DG/CPR/91/326*	IO/T/ENG	J13312	Management of the transfer and diffusion of technology in China's Machine Tool Industry (CMTI)

Project Number	Backstopping responsibility	All.Acc. Code	Project title
DP/CPR/85/017**	IO/T/ENG	J13313	Strengthening of CAD/CAM Centre
DP/CPR/85/087*	IO/T/ENG	J13131	Qualification and Surveillance Laboratory for consumer electronic products
DG/CPR/91/321*	IO/T/ENG	J13313	Computer numerical control system development
DP/CPR/87/025**	IO/T/ENG	J13314	Assistance to the Beijing Household Electric Appliances Research Institute (BHEARI)
DP/CPR/85/019**	IO/T/ENG	J13314	Locomotive test stand
DP/CPR/85/008**	IO/T/ENG	J13316	Technical development of precision die design and manufacture
DP/CPR/85/015*	IO/T/ENG	J13316	China national technical development centre for gears
DG/CPR/91/327*	IO/T/ENG	J13316	Service and developmet centre of fine blanking technology
US/CPR/91/017	IO/T/ENG	J13320	Establishment of the China International Packaging Information Centre (CIPIC) (phase I)
DG/CPR/88/080	IO/T/ENG	J13300	Umbrella project in support of CPR/89/035 - development of turrets for NC lathes
DG/CPR/84/003*	IO/T/CHEM	J13420	Development and application of carbon fibre and composites
DP/CPR/84/004*	IO/T/CHEM	J13420	Research and development in dye-stuffs (phase II) (continuation of DP/CPR/80/061)
DP/CPR/85/005*	IO/T/CHEM	J13420	Beijing Speciality Gas Research and Development Centre
DP/CPR/85/013*	IO/T/CHEM	J13420	Assistance to the Beijing Institute of Chemical Reagents (BICR)
DP/CPR/85/016**	IO/T/CHEM	J13420	Strenghtening the Research Institute of Synthetic Material Ageing (RISMAG), Guangzhou
DP/CPR/85/022**	IO/T/CHEM	J13420	Development and application of silicones
US/CPR/86/130*	IO/T/CHEM	J13420	Recycling system for plastics waste (follow-up of US/CPR/83/245)
US/CPR/87/192**	IO/T/CHEM	J13420	Assistance to Fujian Prosthetics Factory
DP/CPR/88/039*	IO/T/CHEM	J13420	Research and development in castor oil processing
DP/CPR/88/046**	IO/T/CHEM	J13420	Research, development and industrial support for detergent production
DP/CPR/88/047*	IO/T/CHEM	J13420	Application of irradiation technology to reclaim butyl rubber
DP/CPR/88/066*	IO/T/CHEM	J13420	Engineering plastics alloying and filling
DG/CPR/88/070*	IO/T/CHEM	J13420	Umbrella project
DP/CPR/89/001*	IO/T/CHEM	J13420	Development and application of coating technology
US/CPR/90/252	IO/T/CHEM	J13420	International conference on plastics waste recycling technology, Shanghai, China, 15-18 April 1991 (related to US/CPR/86/130)
US/CPR/91/071	IO/T/CHEM	J13420	Technical assistance for the development of the gas industry in China

Project Number	Backstopping responsibility	All.Acc. Code	Project title
DP/CPR/85/004**	IO/T/CHEM	J13421	Development of new technologies for phosphate enrichment for the fertilizer industry
DP/CPR/85/012*	IO/T/CHEM	J13421	Compound fertilizer technology centre (Associated Agency : FAO)
DP/CPR/88/001*	IO/T/CHEM	J13422	Enzyme products development
DP/CPR/89/021*	IO/T/CHEM	J13422	Improved production of penicillin
US/CPR/81/171**	IO/T/CHEM	J13424	Industrial biogas technology demonstration plant and experimental station, Beijing, China
DG/CPR/85/031	IO/T/CHEM	J13424	Coal-water mixture preparation and combustion technology
DP/CPR/85/086*	IO/T/CHEM	J13424	Technology development of water treatment for industrial use in Shanghai
US/CPR/86/093*	IO/T/CHEM	J13424	Assistance with the design and operation of waste water treatment facilities for Beijing, China (phase II - training)
DP/CPR/80/008*	IO/T/CHEM	J13426	Research and development in pesticides (see also US/CPR/83/286. US/CPR/90/026)
US/CPR/83/286*	IO/T/CHEM	J13426	Research and development in pesticides (multifund to DP/CPR/80/008)
US/CPR/90/026	IO/T/CHEM	J13426	Seminar on recent development in the field of pesticides and their application to pest control in China and other developing countries, Shenyang, China, 8-12 October 1990 (related to DP/CPR/80/008)
DP/CPR/85/060**	IO/T/CHEM	J13428	Technical development of a building and sanitary ceramics centre
DP/CPR/85/063*	IO/T/CHEM	J13428	Non-Metallic Minerals Development Centre
DP/CPR/85/064*	IO/T/CHEM	J13428	Research and development for improving building glass technology
DP/CPR/85/088*	IO/T/CHEM	J13428	Assistance to the Research and Development Centre for Light Building Materials
DP/CPR/86/007*	IO/T/CHEM	J13428	Research and development for fly ash utilization (phase II) (continuation of DP/CPR/81/026)
DP/CPR/88/009**	IO/T/CHEM	J13428	Fire prevention technology for high rise buildings
DP/CPR/88/023*	IO/T/CHEM	J13428	China development centre for wall and roof materials
US/CPR/88/171	IO/OS/FEAS	J14101	Pre-feasibility study - establishment of a processing plant for the dehydration and freezing of vegetables (in co-operation with IO/T/AGRO)
DP/CPR/87/024*	IO/OS/FEAS	J14102	Economic evaluation of investment projects
TF/CPR/89/002	PPD ICFM/COOP/NGO	E05100	Strengthening the capabilities of Chinese enterprises for international technical co-operation
TF/CPR/89/003*	PPD ICFM/COOP/NGO	E05100	Promotion of industrial co-operation between Chinese enterprises and partners from Denmark
TF/CPR/90/002*	PPD ICFM/COOP/NGO	E05100	Promotion of industrial co-operation between enterprises in the People's Republic of China and the Republic of Korea

Project Number	Backstopping responsibility	All.Acc. Code	Project title
US/CPR/88/134	PPD ICFM/COOP/NGO	E05101	Development of electronic data communication system
UC/CPR/87/277	PPD ICFM/COOP/STF	E05200	Joint mission for studying possibilities and modalities of co-operation among UNIDO, China and Egypt in the field of phosphate and rare earth production
US/CPR/88/080*	IPCT II/IPAP	G01400	China/UNIDO/German Federal Republic investment promotion programme 1989, Hannover, Germany, 31 March-14 April 1989

* Large-scale project (= total allotment $150,000 or above)

** Total allotment <u>$1 million or above</u>

INDUSTRIAL MAP OF CHINA

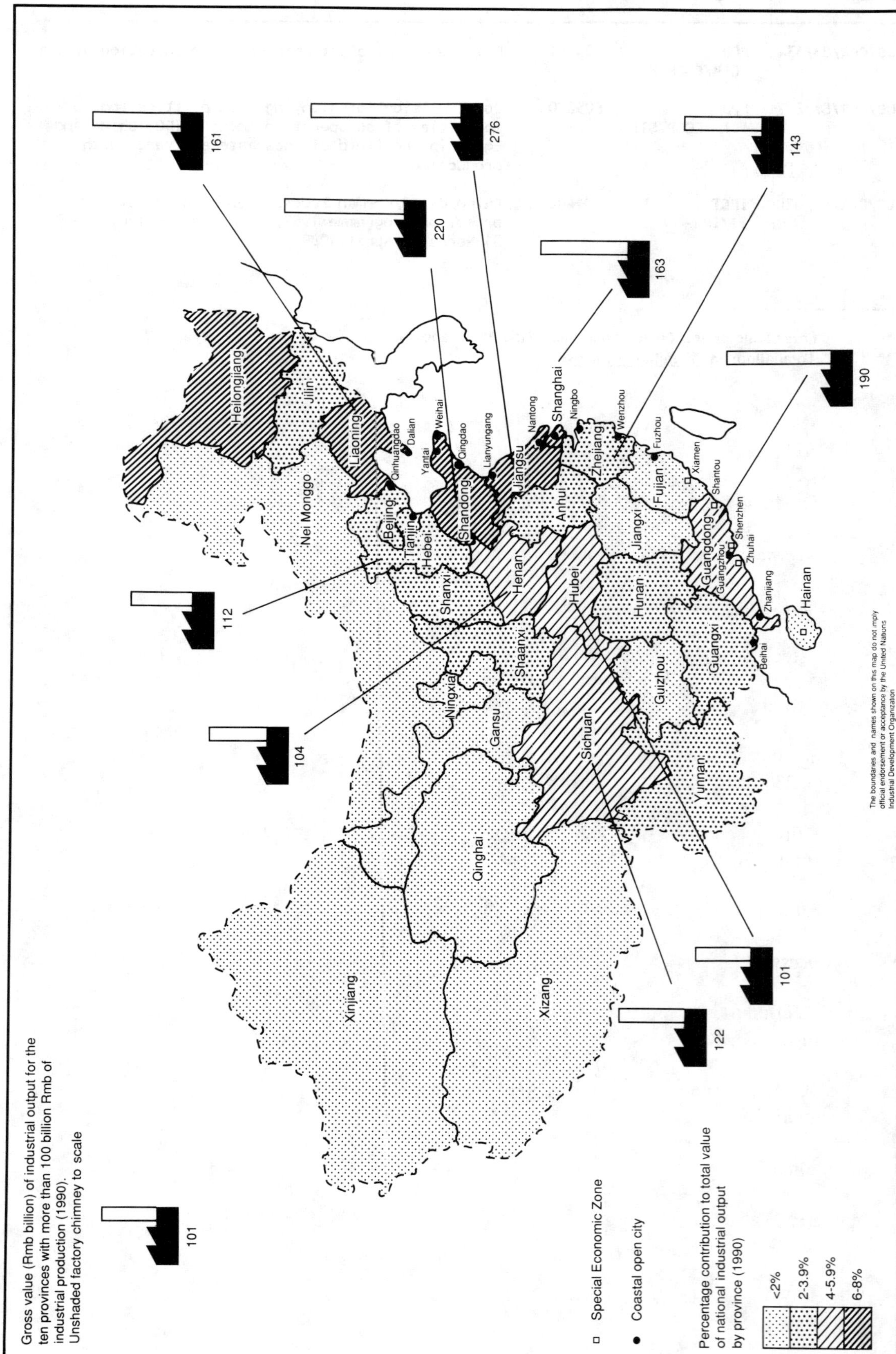

Map prepared for UNIDO by the Cartography Dept., School of Geography, University of Oxford

July 1991

SELECTED REFERENCES

A. Yeats, "China's recent export performance", *Development and Change*, 1984, Vol. 15, No. 1, pp. 1-22.

Beijing Review (various issues).

Bruce Stone, *Fertilizer Use Development in China During the Reform Period*, paper prepared for the World Bank, FAO, UNIDO Industry Working Group, (May 21-23, 1990).

Business China (various issues).

China Council for the Promotion of International Trade and China Chamber of International Commerce, *Trade Promotion* (various issues).

China Daily (various issues)

China Economic Yearbook (various issues).

China Food Industry Almanac, 1989.

China: International Trade Annual Statistical Supplement (various issues).

China Machinery Industries Yearbook (various issues).

China Rural Statistical Yearbook (various issues).

"Chinese steel", *MBM* (December 1988), p. 63.

"Daimler to make heavy trucks in China", *Financial Times*, (October 28, 1988).

Dali Yang, "Patterns of China's regional development strategy", *The China Quarterly* (Oxford, June 1990), pp. 230-257.

David G. Fridley and James P. Darian, "China's resource crisis", *Far Eastern Economic Review* (27 September 1990), p. 62.

Fertilizer International (January 1990), No. 281.

Food processing (August 1988), pp. 112-114.

Gary H. Jefferson, "Potential sources of productivity growth within Chinese industry", *World Development* (Great Britain 1989), Vol. 17, No. 1, pp. 45-47.

Guang Xi Jin, "Productivity improvement in Chinese iron and steel industry", *Steel Times International* (May 1988), p. 52.

International Monetary Fund, *International Financial Statistics* (various issues).

Kenneth R, Walker, "Forty years on: Provincial contrasts in China's rural economic development", The China Quarterly (September 1989), p. 451.

Louise do Rosario, "Asia's fifth dragon", *Far Eastern Economic Review* (8 December 1988), Vol. 142, No. 49, p. 62.

Marc Ross and Liu Feng, "The energy efficiency of the steel industry of China", *Energy* (Great Britain 1991), Vol. 16, No. 5, pp.833-849.

"Nuclear technology boosts agriculture", *Beijing Review* (September 17-23, 1990).

Oil and Gas Journal (January 14, 1991).

Petra Pissula and Dieter Loesch, "Special Economic Zones in the People's Republic of China", *Intereconomics* (September/October 1990), pp. 257-264.

Renmin Ribao (19/4/1991).

Shanghai Industrial and Economic Journal (8 July 1988), quoted in *The China Quarterly* (June 1990).

South (April 1989), p. 37.

State Statistical Bureau of China, *China Statistical Yearbook* (various issues).

Tai Ming Cheung, "Bale of tears", *Far Eastern Economic Review* (22 February 1990), p. 66.

Textile Asia (various issues).

Textile Horizons (February 1990), p. 22.

"The industrial economy", *Beijing Review* (3-8 October 1988), p. 23.

UNIDO, "The People's Republic of China", *Industrial Development Review Series* (Vienna, 2 December 1985, UNIDO/IS.582).

Wen Tianshen, "China's computer industry on the rise", *China Today* (February 1, 1991). pp. 40-42.

World Bank, *China: External Trade and Capital* (Washington 1988), pp. 256-258.

World Bank, *World Debt Tables* (various issues).

World Bank, *World Development Report 1990* (Washington D.C.), pp. 188-189.

World Chemical Industry Yearbook, *China Chemical Industry* (1988 and 1990 editions).

Zhang Dengyi, "China" in *Proceedings of the Meeting of Senior Officials on Human Resources Development in the ESCAP Region*, Bangkok, 17-19 January 1989, United Nations Economic and Social Commission for Asia and the Pacific, pp. 147-168.

Zhang S. and Zhang A., "The present management environment in China's industrial enterprises" in B.L. Reynolds (ed.) *Reform in China* (New York 1987), pp. 52-56.

INDUSTRIAL DEVELOPMENT REVIEW SERIES ISSUED SINCE 1986

NON-SALES PUBLICATIONS[1/]

Country	Subtitle	Symbol	Year
Sri Lanka		UNIDO/IS.613	1986
Cuba		UNIDO/IS.615	1986
United Republic of Tanzania		UNIDO/IS.628	1986
Egypt		UNIDO/IS.637	1986
Mali*		UNIDO/IS.640	1986
Zaire*		UNIDO/IS.644	1986
Pacific Island States: Papua New Guinea, Fiji, Solomon Islands, Western Samoa, Vanuatu, Tonga, Kiribati, The Federated States of Micronesia and Micro States		UNIDO/IS.645	1986
Côte d'Ivoire*		PPD.6	1986
Saudi Arabia		PPD.7	1986
Congo*		PPD.10	1986
Central African Republic*		PPD.11	1986
Colombia		PPD.16	1986
Ghana		PPD.18	1986
The Republic of Korea		PPD.29	1987
Botswana		PPD.37	1987
The Caribbean Region: Jamaica, Trinidad and Tobago, Guyana, Barbados, Netherlands Antilles, Bahamas, Belize, Bermuda, St. Lucia, St. Vincent and the Grenadines, Grenada, Antigua and Barbuda, Dominica, St. Christopher-Nevis, Cayman Islands, British Virgin Islands, Montserrat, Turks and Caicos Islands, and Anguilla		PPD.51	1987
Malawi		PPD.58	1987
Indonesia:	Changing Industrial Priorities	PPD.60	1987
Zimbabwe:		PPD.63	1987
Burma:	Transition to agro-based industrial economy	PPD.65	1987
Jordan:	Stimulating manufacturing employment and exports	PPD.67	1987
Liberia:	Resource-based industrialization and rehabilitation	PPD.74	1988
Qatar:	Towards industrial diversification of an oil-based economy	PPD.75	1988
Nepal:	Industrialization, international linkages and basic needs	PPD.79	1988
Kenya:	Sustaining industrial growth through restructuring and integration	PPD.85	1988
Somalia:	Industrial revitalization through privatization	PPD.91	1988
Philippines:	Sustaining industrial recovery through privatization and foreign investment	PPD.92/Rev.1	1988
Nigeria:	Industrial restructuring through policy reform	PPD.100	1988

Djibouti:	Economic diversification through industrialization	PPD.111	1989
Bangladesh:	Strengthening the indigenous base for industrial growth	PPD.114	1989
Mauritania:	Industrial reorientation and rejuvenation	PPD.115	1989
People's Democratic Republic of Yemen:	Enhancing industrial productive capacity	PPD.122	1989
Yemen Arab Republic:	Diversifying the industrial base	PPD.130/Rev.1	1989
The Sudan:	Towards industrial revitalization	PPD.132	1989
Cameroon:	Coping with reduced oil revenue	PPD.146(SPEC)	1990
Poland:	Second Investment Forum for the Promotion of Foreign Investment (Sales Publications forthcoming in 1991)	PPD/R.36**	1990
Namibia:	Industrial development at Independence	PPD.166	1990
Angola:	Economic reconstruction and rehabilitation	PPD.172	1990
Viet Nam:	Industrial policy reform and international co-operation	PPD.184	1991
Ethiopia:	New directions of industrial policy	PPD.185	1991

SALES PUBLICATIONS[2/]

India:	New dimensions of industrial growth	1990
Pakistan:	Towards industrial liberalization and revitalization	1990
Malaysia:	Sustaining the industrial investment momentum	1991
Poland:	Managing the transition to a market economy	1991

* Available also in French.
** Restricted

1/ Copies of non-sales publications may be obtained directly from: UNIDO, Regional and Country Studies Branch, P.O. Box 300, A-1400 Vienna, Austria. FAX: (0222) 232156

2/ Sales publications may be ordered directly from: Basil Blackwell Publishers, 108 Cowley Road, Oxford, United Kingdom. FAX: (0865) 791347

INDEX